MADE BY NATURE

THE POETRY OF THE BEE

IRIS ROMBOUTS

THIS BOOK WAS CREATED WITH THE SUPPORT OF BIOBEST GROUP

CONTENTS

08	JEROEN OLYSLAEGERS
12	MAURICE MAETERLINCK
24	LANDSCAPE 01
26	LOUIS DE CORDIER
38	THE BEAUTY OF POLLINATION
42	JAN FABRE
48	CARLA AROCHA & STÉPHANE SCHRAENEN
56	BEEKEEPER FASHION
74	LANDSCAPE 02
76	TOMÁŠ LIBERTINY
84	JOSEPH BEUYS
92	LANDSCAPE 03
94	MANCHESTER - CITY OF BEES
112	BEAUTIFUL HONEY BODY
124	LANDSCAPE 04
126	THE POETRY OF THE BEE
138	DAYS OF WINE AND HONEY
150	KOEN ROGGEN
152	ANDREW WRIGHT
162	LANDSCAPE 05
164	MASTER BUILDERS
180	MICHAEL VAN LENT
190	BLOSSOM TIME
200	RECIPES
202	AND FINALLY...

BUZZ BUZZ BUZZ BEE ACTIVE BY JEROEN OLYSLAEGERS

Bees are eternally buzzing in my head. In February 2012 I write my first column about the importance of the bee, in the newspaper De Morgen. About a year later, in March 2013, I publish an appeal in the same paper to send a mail to the Minister of Agriculture of the time, begging him to follow the EU advice and temporarily ban the use of neonicotinoids in pesticides. There is a strong indication that these substances change the bees into 'zombees', making them completely disoriented and preventing them from functioning normally. The battle against these neonics takes years of patience and perseverance by the activists. While I write this text, in the summer of 2018, the Belgian government tells the EU that it wants three years deferment of the ban on neonicotinoids. The reason? National agriculture isn't ready for it. That says a lot, a hell of a lot. The bee and all things concerning the bee is the canary in our 21st century coalmine. For many decades, Belgian agricultural policy has been a subsidised, industrialised, depressing, vastly expensive disaster, exploiting the farmers. The reason why neonics are still being used here is because there isn't enough organic farming and also because laziness prevails. This is actually so much as being admitted, because it seems there are 'no alternatives' and our so-called politicians need another three years scratching their lazy butts, doing what they're best at: leaving things alone. It's a filthy shame.
Good, got that off my chest.

When I started to write about the honeybee becoming extinct, I was not alone. I think that awareness about the precarious fate of the honeybee started to peak around 2011 on all sorts of ecological forums and that this is how it reached mainstream media. The reactions to my writings were rather weird. Jeroen worrying about the bees, how cute. Further on in this exceptional book you can read about the link between our sole Noble Prize winner Maurice Maeterlinck and the bees. I once took part in a literary quiz where a bit was read from his 'Vie des Abeilles' and we were asked who the author was. Apart from myself, nobody knew the answer, but it got a good laugh. Bees are touching at their best, but mostly ridiculously small and unimportant, the reactions implied. Of course, the opposite is true: the fates of humans and bees are even more intrinsically linked than the incredible obsession of some Western people to eat meat twice day.
By the way, I'm writing this text in the home of Cath, a friend who has just become a beekeeper. Her boyfriend showed me pictures of her in her white beekeeper's hat, her protective clothing, and it reminded me of the wonderful drawing by Pieter Bruegel the Elder. 'The Beekeepers' is from 1568 and like the vast majority of his work it has more than one meaning. In the foreground, we see three beekeepers, protected by long sleeves and a basket they wear as a mask. It is not an innocent scene; it is unlike Cath taking care of her beehive. To the right, someone is climbing a tree, probably aiming for the eggs in a nest, and the three masked men themselves seem to be stealing the honey instead of really taking care of the bees. Hardly a year later, Bruegel was dying and he ordered his wife to destroy several drawings, probably because the scenes he had drawn were too critical of the bloody reign of the Duke of Alva, who, at the time, mercilessly imposed his papal truth on the Southern Netherlands. 'The Beekeepers' may have escaped the attention of Bruegel's wife, or maybe it wasn't seen as confrontational enough to cause her problems. Yes it is not hard to see it as an accusation against thieves who grab whatever they can get, instead of being real leaders who tailor their goals to long-term thinking; this is stealing instead of governing.

It has always been this way. Bees and beekeepers have always spoken to the imagination of humankind. In the past, several political and social thinkers have used the beehive as a symbol: both liberals and socialists and many representatives of other tendencies applaud the organisation in the bee world as a political model. The beehive can stand for ultimate human utopia - or its exact opposite. In the wonderful book 'Honeybee Democracy' (2010) Tom Seeley explains the science behind the model of the honeybees. In late spring or early summer, honeybees like to swarm. Seeley explains that a demographic process takes place amongst the bees, about where exactly to go. They take the decision based on the spots their scouts have discovered; they communicate their findings via the so-called 'bee dance'. Scientific research about this was only started some seventy years ago. Did you know that a certain Karl von Frisch from Austria only discovered in the summer of 1944 that a worker bee can explain to her congeners where a food source is located through a certain dance? In 1973 he was awarded the Nobel Prize for his findings, together with his compatriot Konrad Lorenz and Dutchman Nikolaas Tinbergen. Since the ideological bloodlust of the twentieth century, we are quite wary about group intelligence, but Seely's book makes it clear that in the bee world, it is perfectly possible to take the right decisions as a big group. While I'm typing this, I am still amazed myself. Since the book appeared, I have learned that humankind are a bunch of bunglers compared to the Apis mellifera, the honeybee and the democratic principles of its kind, with long-term thinking as a goal.

How ironic, therefore, that the existence of this species (and, of course, many other insects), is threated by humankind's short-term thinking, the same thinking that Bruegel subtly denounced. That is the dark side of our relationship with bees, and its name is, in fact, indifference. On the other hand there is something I could call 'care' or 'taking care of'. Writer and artist Dirk Hughes movingly describes the advantages of beekeeping in a TED-talk, calling it 'peace of mind to know that one plays an active role in the care for a wondrous species on this spectacular planet.'

It's all a little abstract to you and me, unless you practice beekeeping. I suspect that my friend Cath, the brand new beekeeper, understands this, having given me books about the importance of bees. A communal friend has an uncle who works very intensely with bees, and we go and visit him together.

On a Wednesday afternoon we suddenly stand in the wonderful garden of 'uncle Guido' a human being with clear eyes and a white beard. His real wealth lies behind the trees and bushes, in a part of the garden his wife never visits because she is allergic to bee stings. I am amazed at what we find there. We are surrounded by homemade beehives and by the buzzing of many bees in the shade of this beautiful summer's day. There are fifty hives. More than two million honeybees… None of us wears protective headgear; Guido spent many years creating a fairly calm bee species. As soon as he knows that Cath has become a beekeeper, he starts telling us about his hobby which has got quite out of hand. Someone from our group asks how much honey it gives him. But he doesn't care about the honey. He has somebody who comes and helps him, but in fact those hundreds of litres of honey are not the point. Guido artificially inseminates queen bees and then sells them. What is so fascinating is that he is also engaged in a project with Ghent University fighting the varroa mite, a much-feared enemy of the bee population. Their weapon is called thymol, a chemical substance found in the herb thyme. Thymol attacks the mite's airways and also allows the bees to discover early on whether their larvae have been affected. Now that I think of it, the Egyptians used thyme to embalm their dead and for many years it was the precursor to the antibiotics we use nowadays. It was also used to disinfect medical bandages in the old days.

Guido shows us his workshop. He makes everything himself. He even lived his own bee apocalypse, a few years ago. During the winter, they were using heavy drills and diggers on the land next to his. This disturbance of the wintery silence cost forty-seven out of his fifty colonies their life. But after a year, the hives were full again. Here, a stubborn love reigns, an attention to detail, an everlasting search to understand the Apis mellifera in order to enhance and improve the relationship between human and insect. And all this not in a lab, but in a large garden behind a semi-detached house.

Guido starts to tell us about genetic copies, the possible threat of the Asian hornet the beekeepers may have to face, the dances executed by the bees and their various languages. "In fact," he tells us, "the different scents the bees excrete are also a communal language." I ask several questions and if I have understood correctly, it seems that bees even send each other 'scented objects', three-dimensional holograms that consist of various scents, in the way that dolphins send each other resonant signals that form images. There I am, amongst so many beehives, next to someone who has dedicated much of his life to the honeybee. I take photos, I use my smartphone to film Guido's explanation when he extracts and studies honeycombs, shows us this and that, and I am stunned. I can imagine that nature-lovers frown upon the artificial insemination of queen bees, which is what Guido the 'queen maker' does. There are vegan beekeepers who handle their bees in a totally different way. This could lead to interesting discussions, but what became clear to me, there amongst the beehives, is how strong the bond between human and bee can become and how intelligent their mutual collaboration is.

This book, reader, is about the special relationship between man and bee. I am glad it has been published, because it was born out of care, fascination and awareness. We humans are strange creatures. Our moral superiority towards everything that lives on our planet is our weakest characteristic. It is ridiculous and yet tragic, because we pay a very high price for it. The only thing that can save us is awareness, and this book helps us and points us in the right direction. Finally, at the end of the day, only one thing counts: to take care. Maybe we can start by reading this book.

It is a gift.

And I would love to be a bee-keeper.

You too?

They make honey, we make poetry.

May bees buzz around your head forever. ✖

A NOBLE PRIZE WINNER & HIS BEES

MAURICE MAETERLINCK

Maurice Maeterlinck (Gent, 29 August 1862 - Nice, 6 May 1949) is a Flemish author who wrote in French.
In 1911 he was awarded the Nobel Prize for literature 'in recognition of his wide-ranging literary activities and in particular for his plays which distinguish themselves by an abundance of imagination and poetic refinement. These, often in the guise of a fairy tale, are a great source of inspiration, at the same time mysteriously appealing to the feelings of the reader and stimulating his imagination,' according to the jury report.

Maeterlinck was raised in a well-off family. He studied at Ghent University, at a time when all teaching was in French, and became an attorney in 1885. He did speak Dutch and even pleaded in that language, which was very unusual at the time. However, he didn't do it for long, because he soon completely devoted himself to literature. He was a good friend of author Cyriel Buysse and they always conversed in the Ghent dialect.

Author Marnix Gijsen, who worked for the government as Belgian Commissioner for Information in New York and who was said to speak English with an Antwerp accent, stated that Maeterlinck's French had a Ghent sound to it.

In 1939 he was appointed Grand Officer in the Order of Saint James of the Sword. Maeterlinck mostly wrote plays, but also poetry and a number of essays. In 1901 he published 'Het leven der bijen' (the life of the bees), a work of more than 200 pages, stylistically divided into seven books. Before he wrote it, he had extensively studied the life and habits of these insects. He himself owned and looked after three hives, and conducted numerous experiments to witness the reactions of the bees for himself.

This is an introduction to his analyses, dotted with various quotes from his book, which can, without exaggeration, be called his 'bee bible'.

The First Book is titled 'On the Threshold of the Beehive'. It is a general introduction and the author begins with: *'It is not my intention to write an essay about bees and beekeeping'* and *'I will hardly write anything that is not already known to all people who have dealt with bees in any way.'*

So what is his intention? He simply wants to talk about bees, the way someone talks about any subject he knows and loves. To justify this decision he mentions all texts - and there are many - that have been written about these wonderful insects, from ancient Greek times to the 20th century. He claims to have read them all and they were either too mythical or too scientific, making them too mysterious or too complicated for the average reader.

He gives a lyrical account of the first time he saw an apiary, writing in his beautiful descriptive prose: *'It was years ago, in a large village in the charming, tidy Dutch Flanders, which more than Zeeland itself concentrates a feeling for bright colours, like a hollow mirror of Holland, it is a feast for the eyes as well as a beautiful, sturdy toy, with its colourful façades, towers and carts, its shiny cupboards and clocks in the corridors, its trees standing side by side along quays and canals, as if waiting for some pleasant, simple ceremony, its boats and barques with decorated sterns, its doors and windows like flowers, its neat locks, its well-finished, colourful drawbridges, little houses, varnished like shiny pottery, from which women in bell-like clothes adorned with gold and silver appear, to milk the cows in fields surrounded by white fences, or to spread the linen on the flower-filled green grass, divided into ovals and squares.'*

Talk about a sentence rich in details....

He tells us that the year of the bee goes from April to the end of September and before we open the bee hive: *'it is sufficient to know that it consists of a queen, the mother of her entire colony, of thou-*

sands of worker bees and genderless, undeveloped and infertile females, and of a few hundred drones, from which the sole unhappy husband of the future ruler will be chosen, the one that will be elected to be queen after the more or less voluntary departure of the ruling mother.'

But be careful: someone who doesn't know the character and habits of bees and opens the hive anyway, will discover that it changes straight away into 'a burning bush of ire and heroism.'

He talks extensively of the danger of bee stings: *'that cause such a very peculiar pain that it is almost impossible to know what to compare it with: a flashing dryness one could say, a kind of desert flame that spreads over the wounded body part; it is as if our daughters of the sun had pulled a splendid venom from the heated rays of its origin, in order to even better be able to defend the treasures of sweetness that they owe to the benevolent hours of the sun.'*

Moreover he talks about the observations of the uninitiated when opening an observation hive. That person is disappointed because he doesn't understand what he sees.

He doesn't know enough: *'to fathom the almost perfect but ruthless society of our hives step by step, a society where the individual is completely swallowed up by the republic and where the republic in its turn is sacrificed to the abstract and immortal state of the future, following set rules.'*

At the end of 'The First Book' he claims that the hymenoptera, which is the name of the Northern bees, are the most privileged inhabitants of our earth after humans where intelligence and development are concerned: *'however, it is shown that the improvement of the species can only be achieved at the expense of the freedom, the rights and the happiness of the individual himself. As society becomes organised and evolves, the individual life of each of its members loses territory. As soon as improvement can be seen anywhere, it is only the consequence of a more complete sacrifice of what is personal to the communal interest.'*

He will prove these ideas in the next 'books'.

The Second Book is entitled 'The Swarm'. As early as February, the winter lethargy is over in the hive. The queen starts laying eggs again and in springtime, the worker bees have already visited the anemones, the danewort, the stinging broom, the violets, the willows and the chestnut trees. *'The attics and cellars overflow with honey and pollen and thousands of bees are born every day. The abundance in the all too prosperous colony becomes such that hundreds of worker bees that are slightly late coming back from the flowers in the evening, don't find a space anymore and are forced to spend the night in front of the entrance, dying with cold.'* Soon, there is unrest in the hive and the old queen feels that something new is coming. She will have to leave the colony that she reigns over. *'She is not the queen in the sense we humans consider the queen. She gives no orders and, just like her subjects, she depends on a hidden all-knowing power that, while trying to detect it, we will call, for the time being, 'the spirit of the hive'.'*

Then the tasks of this 'spirit' are described, and they turn out to be extensive and *'ruthless but with modesty and seemingly serving some higher duty.'* The spirit organises the work of everything that lives in the hive and also determines when swarming should start, *'a whole colony, at the height of its success and power, suddenly leaves all its treasures, palaces, homes and the fruit of its labour to the new generation, and goes to find the insecurity and emptiness of a new homeland... Obeying a higher law than the happiness of that colony....'.* Of the 80,000 to 90,000 bees of the whole colony, some 60,000 to 70,000 leave the hive, at the very moment when it has reached its pinnacle of prosperity.

The author then gives an extensive description of what the hive looks like at that time and how the bees communicate, and he concludes that swarming is not a blind exodus, but a well-considered sacrifice of the current generation for the future one. Once again he points out that it is not the queen, but the 'spirit of the hive' that decides on the swarming. He then describes the hustle and bustle and the chaos in the hive during the preparation, and how they influence the behaviour of the queen. Her relationship to the worker bees is described extensively and we find out that: *'every maggot of a worker bee that is not yet three days old can, by means of special food, be transformed into a royal nymph.'* He calls this the greatest democratic principle of the beehive. The little head of the worker bee, moreover, contains the convolutions of the greatest, most perfect brains in the hive: *'In fact, apart from the human brain and in a different order and organisation, it is the most beautiful, most tender and most perfect brain in nature'.* After that, he gives a poetic description of the mass exodus from the hive, and how the swarm moves as one whole until the queen descends on a branch, where she is immediately surrounded by a cluster of thousands of bees that wait patiently for the return of the scout bees that are looking for a new refuge.

If there is a beekeeper nearby, he can easily lure a whole swarm into an empty hive. If not, the whole cluster discusses the suggestions of the scout bees and that can take several days. Then they depart, like a trembling cloud, straight to a fixed and always far-away destination. They return to nature and the humans lose their trail completely.

The Third Book is entitled 'The Foundation of the City'.

It describes the first activities of the swarm when it has landed in an empty hive. The worker bees have no time to spare because the queen wants to

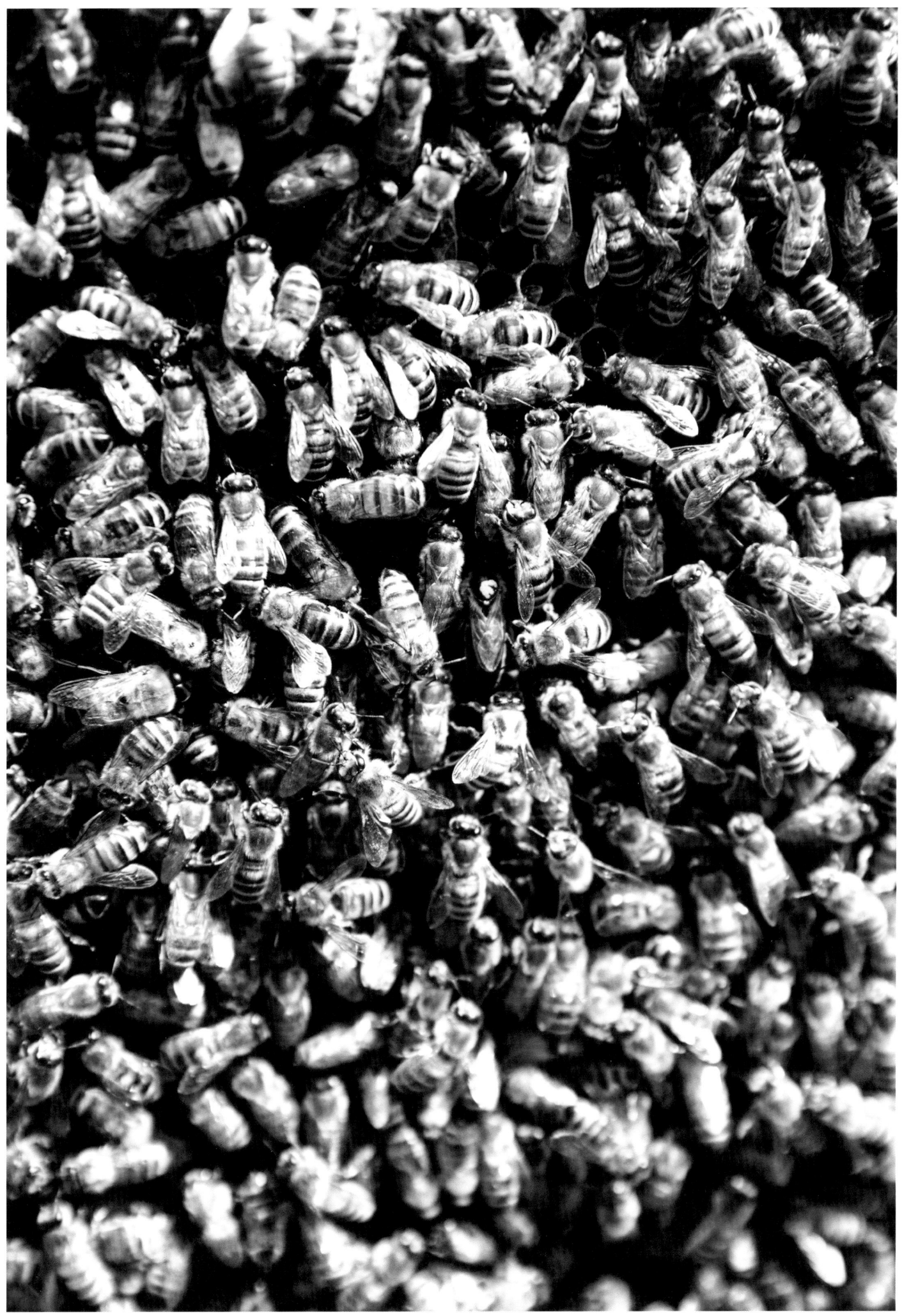

The queen bee is marked with a coloured dot | 17

start laying eggs straight away, and is already spreading them on the ground. If the beekeeper has provided a hive in which various windows are hung on top of each other, they continue to work on this artificial honeycomb and: *'in less than a week they own a city that is just as abundant and well built as the one they left behind, while if they had had to do it all themselves, it would have taken two or three months to build the same number of storage rooms and homes in white wax.'*

In this 'book' the author also mentions some peculiar characteristics of the bees. *'Apart from their miraculous industry, politics and selflessness, one thing will always surprise us and rein in our admiration: their indifference to the death and suffering of their fellows. There is a strange contradiction in the character of bees. In the hive they all love and help each other. There, they are one, like the good thoughts of a single soul. If you hurt one of them, thousands will sacrifice themselves to revenge this insult. But outside the hive, they don't want to know each other'.* When, through certain circumstances, 10 or 20 bees from the same hive are crushed: *'the others ignore them and calmly continue to use their tongues, beautifully shaped like a Chinese weapon, to draw the liquid that is worth more than their lives, and remain indifferent to the death throes, ignoring the convulsions and cries of despair around them'.*

About the way they pass on messages, he writes: *'It is obvious that they understand each other, and a republic with so many members, whose tasks are so different and yet so wonderfully complementary, could not exist if so many thousand of creatures lived together silently and in spiritual separation. Therefore, they must have the ability to express their thoughts and feelings, either through a vocabulary of sounds, or by means of a tactile language or magnetic intuition that may even be located in the mysterious antennae that feel and understand the dark and that in worker bees consist of approximately 12,000 feeler hairs and 5,000 olfactory cavities'.*

Every day, orders are given in some way or another: one day, worker bees have to collect only pollen, the next day, only honey juice. The conditions surrounding the hive are also reported by the scout bees every morning: *'today the lime trees by the canal are blossoming, the white clover glitters in the grass along the road, the lilies and mignonettes are overflowing with pollen'.* Then measures are taken to divide the work.

But let's go back to the building of the new city. A chain of countless bees attaches itself to the top of the dome and, stuck together, they form an upside-down cone that appears to be able to turn honey into wax. This wax is used to continue the build. Bees always build four different kinds of cells: the royal cells that resemble acorns, the larger cells in which the drones are raised and where provisions are stored, the small cells that are used as the cradle of the worker bees and common storage space, which take up about eight tenths of the built-up surface in the hives, and finally the transition cells that connect the large ones with the small ones.

'Each of these cells is a hexagonal tube resting on a pyramidal basis, and each honeycomb consists of two layers of these tubes, which rise up from the basis towards the opposite side.'

The honey is collected in these tubes. The final result looks like this: *'the hive is sectioned from top to bottom by five, six, seven, eight and sometimes ten segments that are completely parallel and that resemble large slices of bread, descending from the top of the bell-jar and adapting themselves flawlessly to the sides of the jar. Between each of these honeycombs is an 11 millimetre space where the bees live and walk around'.* When the royal cells are finished, the queen moves in with her following of guards, confidants and servants. She then begins to lay her eggs and doesn't stop until the first autumn cold sets in.

The Fourth Book is all about 'The Young Queens'.

The young hive described in the Third Book is closed and the former hive is studied to see what happens there after the departure of the swarm. Two thirds of the bees have left for good, but a few thousand remain and they stoically resume their usual duties: keeping the hive clean, visiting the flowers, storing the honey, guarding the storehouse, etc. The mother city no longer has a queen, because she has left with the swarm, but she has left thousands and thousands of eggs behind and they have become white maggots by now. Measures are taken to speed up their transformation, and our author takes us to: *'the chamber of life preceding life. There, in hermetically closed cradles, in the endless connection of wonderful hexagonal cells, myriads of maggots are waiting for the moment of awakening'.* Some hundreds of worker bees are dancing and waving their wings to produce the necessary warmth. After a few days, the lids of these urns start to open and the new-born bees appear. They are cared for, fed and educated immediately, but the various tasks they need to fulfil have been programmed into their brain already. After eight days, the new bees venture out of the hive to perform their 'cleansing flight'. This way, the bee can encourage her air duct to swell, expanding her whole body. Then, she waits another week and then she embarks on her first outing as a flying bee, together with her older sisters. That flight is also a flight of discovery, when she absorbs all the observations that allow her to find the hive without problems for the rest of her days, however far removed she is. Now, the young bee has become an adult.

However, the city that was deserted by the swarm still hasn't got a new queen. But at the sides of one of the middle honeycombs, seven or eight strange structures appear. They are hanging,

From author to author: Tony Rombouts writing about Maurice Maeterlinck and his bees

completely closed acorns that take up the space of three or four worker bee cells. An extremely alert, careful guard watches over this area. In each house, the departed queen has laid an egg. Three days later, a larva appears, and it is fed copious and very special food. This causes the little larva to develop in a totally different way. Once it has shed its pupa, the new queen - if she survives a fight with the competition - will live for four or five years instead of six or seven weeks like the worker bees. Her body will grow twice as long, her colour will be shinier and lighter, her sting will be curved. Her eyes will have only 8,000 to 9,000 facets instead of 13,000. Her brain will also be more limited, but her ovaries are enormous. She will not long for the sun or the open air, and she will die without ever visiting a single flower. She will spend her life in darkness and will untiringly visit uncountable cradles to lay eggs.

The royal nymphs that sleep here are not all the same age and the nurse bees first start to slowly thin out the walls of the ripest cradle, so that the young queen who helps from the inside, can soon leave her wax home. She is immediately cared for and fed, and after some ten minutes she is quite stable on her feet. Nobody has been able to find out why, but she already feels that she is not the only almighty bee and that she needs to conquer her kingdom. That is why she seeks out the other royal cradles and the guards move away to give her access. Torn by furious jealousy, she attacks the first royal cell and uses her teeth and legs to break the wax. If she succeeds, she violently removes the pupa, unrolls the sleeping princess, turns around and stings the competing embryo in the open cell until she dies. All the other royal cells get the same treatment and the surrounding bees do nothing. Afterwards, they remove the stricken nymph and throw it out of the hive. Then, the new queen is accepted as their ruler by all of them.

But there can be a different scenario, especially if the worker bees, encouraged by the spirit of the hive, long to form a second swarm.

In that case, the first-born also approaches the royal cell: *'but instead of finding her humble servants there, she meets an enemy power blocking her way. Excited and driven by her fixed ideas, she want to break through with violence or choose a different way, but everywhere, she is held back by guards looking after the sleeping princesses. She wants to continue her attack, but is pushed back with more and more violence and she is even abused, until she herself begins to vaguely understand that the small worker bees represent a law that overrules the law that drives her. Finally, she goes away and, still furious, she goes from honeycomb to honeycomb, uttering a strange war song or a threatening complaint that sounds like the silver tones of a trumpet. This sound of powerlessness and bitterness is so loud that it can be heard three or four metres away, right through the double walls of the hive. The sound has a magic effect on the worker bees. It fills them with reverent fear or dismay, and the guards surrounding and harassing her suddenly stand still, bow their heads and wait without moving.'*

These plaints can be heard for three, sometimes five days. They are meant to wake the sleeping competitors and call them to battle.

'In the meantime, they are developing, want to see the light and start to gnaw through the lids of their cells. The republic is threatened by chaos. However, when the spirit of the hive took its decision, it foresaw all the consequences of this decision and the guards, who are very well informed, know from hour to hour what they should do to counteract the attacks of the thwarted insect and to make the two enemy forces work together towards the same goal. They know that the young queens who long to be born, would, if they managed to escape, fall into the hand of their invincible sister, who would destroy them one by one. So while one of them gnaws through the walls of her prison from the inside, they cover it with a new layer of wax, and the impatient princess doggedly continues, not knowing that she is up against an enchanted obstacle that rises from the ruins time and again. In the meantime, she hears the challenging cries of her competitor, and because she knows her destination and her royal duties before she even has managed to get an glimpse of life and knows what a hive is, she courageously answers them from her own prison. But since her cries have to penetrate the walls of a grave, they sound muted and hollow.'

In the meantime, the second swarm already stands at the ready and lures the new queen to come with them.

'Immediately after her departure the worker bees that have remained in the hive release one of the prisoners, who then makes the same murderous attempts and utters the same cries of anger, and in most cases they allow this new queen to murder the others'.

Then, normal life continues and the work is done with even more zeal, because the worker bees are all very young and the hive has been depopulated, so that large gaps need to be filled before the winter.

There is a third possibility. In rare cases, it happens that two queens are born at the same time. Then, an often very long fight ensues, with stings and teeth, until one of them is killed. In that case, the guards and nurses never interfere.

The winner is always accepted as the only recognised ruler. Yet, she is still a virgin and she needs to organise her mating flight before she can lay her thousands and thousands of eggs.

Therefore, 'The Mating Flight' is the title of the Fifth Book, and Maeterlinck describes this flight in the most poetic of ways. But first we need to mention the male bees, the drones. In the hive, the virginal queen is surrounded by hundreds of males who are always saturated with honey and have no task to fulfil at all. Only one of them is tasked with impregnating the queen. However, the mating never takes place in the hive. In spite of their 13,000 eyes on both sides of their head, and their antennae, with 8,000 cavities for scent, they do not notice the presence of the queen within the confines of the bee city. They can only discover her existence when she floats in the sky. The queen waits several weeks after her birth until a beautiful morning breaks and then: *'she likes to choose a moment when not much dew wets the leaves and flowers with memories, when the last chill of dawn is still wrestling with the heat of the day and is almost beaten, like a naked virgin in the arms of a strong warrior, when the silence and the roses of the approaching afternoon still allow in a faint scent of morning violets or the transparent cry of the dawn'.*

'She flies backwards, returns two or three times to the entrance and when she has carefully taken in the appearance and the exact location of her realm that she has never seen from the outside, she flies, like a bow from an arrow, to the zenith of the azure. She reaches a height and a stratosphere of light where the rest of the bees never venture in their lives. From afar, the drones have noticed her appearance and breathed in her magnetic scent. Immediately, large groups gather and start to pursue her in a sea of joy, the transparent boundaries of which are constantly pushed back. She, drunk from her flight and obeying the wonderful law of the species that chooses her lover for her and allows only the strongest to catch up with her, in the solitude of the ether, she still ascends, and for the first time, the blue morning sky powerfully penetrates the air holes in her body, and sings like heaven's blood in the thousand ramifications of both air ducts that take up half her body and swell in space. And still, she goes higher. She needs to find a lonely space that is not visited by birds, because they could disturb the mystery. Higher she goes and the disparate group that follows her is getting smaller, and some break loose. The weak, the infirm, give up pursuit and disappear in space. There in the endless opal, a small, tireless group remains. She demands a last lift of her wings and look: the one that was chosen by a mysterious power flies towards her, embraces her, penetrates her and borne by the double speed of their communal flight, the mounting spiral of their embrace lingers for one second in the hostile passion of love'.

Nature, though, is very cruel, because as soon as mating has taken place, the belly of the drone opens, his sexual organ is torn off, taking most of the intestines with it, the wings relax and, as if this marriage had killed him in a flash of lightning, the empty body briefly turns around and then falls into the abyss. The queen, however, lands safely, and removes the remains of her torn-apart lover that still stick to her body. At the entrance of the hive, several groups of bees are waiting for her, and they accompany her to her royal quarters. Two days after mating, she lays her first eggs and from then on she is treated with the utmost care by her subjects. She never leaves the hive again and she will never see light again, unless she has to accompany a swarm, and her fertility ends when death approaches.

In the Sixth Book the most tragic part of the bee year is described: 'The Drone Massacre'.

After the impregnation of the queen the fields are still full of flowers and, thanks to the hard work of the worker bees, the hive thrives. The drones have fulfilled their only task and they are now superfluous, but they are still tolerated for a while. Not only do they lead a life of idleness, they hang around and hinder the busy work of the others. They choose the best corner of the house, they eat the most fragrant honey and they soil the honeycombs with their excrement.

For the time being, the worker bees patiently repair all the damage, but soon, their patience runs out. One morning the spirit of the hive gives a password; it goes from mouth to mouth and the placid worker bees suddenly transform into judges and executioners. 'That day, part of the population does not go looking for honey, but dedicates itself to the work of justice. The fat lazy slobs sitting around in bunches on the honey-carrying walls, just sleeping, are suddenly woken by an army of furious virgins. They awake happy and unknowing, they don't believe their eyes and it is only slowly and with effort that their surprise breaks through their laziness, like a moonbeam through the water of a marsh. They imagine they are victims of a misunderstanding, look around in wonder, and because the basic idea of their life awakens in their confused brains, they take a step towards the honey cells to find sustenance. But they are over, the days of May honey, of the flower nectar of the lime trees, of the easily obtainable ambrosia of sage, thyme, marjoram and white clover. Instead of the free access to that delicious full reservoir that gave up its sweet, wonderful store as soon as their lips touched it, they see themselves surrounded by a burning bush of poisonous stings, aimed at them. The whole atmosphere of the city has changed. The wonderful scent of nectar has been replaced by the sharp scent of venom, hanging in thousands of drops at the tips of the stings, sowing resentment and hatred. Before they have fully realised the incredible reversal in their voluptuous existence, the overthrowing of the wonderful laws of the city means that each of the shocked parasites is attacked by three of four servants of justice, who do their best to cut off his wings, to saw through the thin stem connecting the back half of his body with the front, to remove the trembling antennae, to dislodge the legs and find an opening between the rings of the armour and plant their swords. In spite of their size they have no weapon, they don't have a sting, they do not think about defending themselves and try to get away, or pitch their clumsy mass against the thrust of their attackers. They lie on their back and with strong legs, they clumsily hold on to their enemies, but these don't give in. Soon, the wings of the poor victims have been pierced, their legs have been torn off, their antennae gnawed off and their beautiful black eyes show nothing but despair and fear of the end. Some succumb to their wounds and are immediately carried to a remote burial ground by two or three of their executioners. Others who are not so heavily wounded manage to hide in a corner, where they are heaped together and where a relentless guard keeps them prisoners until they die of lack of food. Yet many manage to reach the entrance and flee outside, but when evening falls, driven by hunger and cold, they return to the entrance of the hive in droves, to beg for shelter. There, the implacable guard awaits them again. The next morning, the worker bees clean up in front of the exit, where the bodies of the drones are piled high, and the memory of these good-for-nothings is erased from the city until next spring.'

After the execution of this death sentence, regular work continues, but it slows down because the honey juice in the flowers becomes more and more rare. The autumn honey is still stocked to complete the necessary food supply and the bees start preparing for hibernation: 'They gather in the centre of the hive, clutch onto each other and hang from the honeycombs that contain the loyal urns, from which in the days of icy cold the distorted substance of the summer will appear. The queen is in the middle, surrounded by her guards. The first row of worker bees clings to the sealed cells, a second row covers them and is, in turn, covered by a third, and so it continues to the final row, which forms the casing. When the outer bees feel overwhelmed by the cold, they return to the deeper layers and others take their place. The hanging mass looks like a reddish-brown, lukewarm ball, cut through by the honey walls, and almost invisibly going up or down as the cells it clings to become emptier. By means of a simultaneous movement of their wings they keep a constant temperature inside the ball, equal to that of a spring day. The bees that are on top of the filled-to-the-brim honey cells, offer the honey to their nearest neighbours, who offer it to others. This way, it goes from leg to leg and from mouth to mouth until it reaches the most remote of the group.'

And so, the hive survives the hard winter until, at the start of the next spring, work is resumed.

In the last book, the seventh, Maeterlinck discusses 'The Progress of the Species'. He admits he still hasn't found 'the spirit of the hive', but in this part he no longer writes about life within the hive, because he has covered everything that happens within a year, and everything repeats itself year after year. Therefore, he now gives an overview of the various kinds of bees that live in different parts of the world. He also tells us more about their history, because bees were already kept by the Ancient Egyptians. In the meantime, the Northern bees that live in our region have learned to understand and accept human interference. Therefore, they prefer a man-made hive to a nest they have to find in nature. They have cleverly taken advantage of the foundations with imprints of cells that the beekeeper provides, so that they don't have to build every cell themselves.

It looks as if, as long as enough flowers bloom, humankind can continue to count on the honey that has been diligently collected by bees. ✖

LOUIS DE CORDIER

PORTRAIT OF THE ARTIST AS AN OLD SOUL

They call him a Homo Universalis. He is often compared to Dürer or Da Vinci. He shrugs. A conversation with Louis De Cordier: artist, philosopher, scientist, archaeologist, beekeeper, a really nice guy and a true Renaissance man.

"I used to think of myself as a traditional artist," De Cordier says. "It's how I started: creating sculptures and installations, organising exhibitions." Yet he is a unique and original artist who refuses to be pigeonholed. His work explores unknown territory, and his art is of a broad and startling scope.

In 2008, his interest in electromagnetic radiation led him to Egypt as the leader of the Mataha expedition. The Belgian-Egyptian team went looking for the long lost labyrinth of Hawara as described by Herodotus and Strabo. "It seemed like a nice reversal of a well-known principle. Instead of directing your antennas to the sky, why not aim them at the ground, to see what's underneath. Of course I knew about the labyrinth, but it was still a huge surprise when the geophysicist said that a gigantic structure could be seen. All of a sudden I was at the centre of this amazing discovery, not really knowing how to categorise it. It was a tempestuous time in my life as an artist, and I could not help but be swept away."

Categorising such a discovery as an artistic endeavour is difficult enough, let alone capitalising on it." De Cordier: "The art world is rather traditional. They prefer a work that looks nice in a gallery. A painting, an abstract sculpture, even a video or an installation. But what on earth are they going to do with this archaeological expedition, an undertaking on the cutting edge of art and science?"

As an artist, De Cordier creates structures and spaces that combine art and architecture with science and nature in the broadest sense. His projects often have a sweeping, almost mythological grandeur to them; cocoons, igloos, antennas. He is currently building, high up in the Spanish Sierra Nevada, an ark containing almost seven thousand books he deems significant for future generations. Each of them has been painstakingly vacuum-packed to prevent them from insect infestation and decay.

The Biblioteca del Sol in Andalusia is filled with dreams, knowledge and human aspiration, with ecology and sustainability at its very core.

De Cordier's vision and creativity might be rooted in the past, but his ideas are firmly focused on the future. He is passionate about preservation, spirituality, self expression and human survival. In his work, De Cordier has often recreated the Platonic and Archimedean Polyhedra (one of the 13 solids first enumerated by Archimedes) that are often prominent in the drawings of celebrated Renaissance artists which were said to 'transport the wisdom of times through their perfect, almost holy geometry'. De Cordier is not afraid that such associations are referenced to him and his work, shrugging off such labels as 'Atlantis seeker' or 'End-of-the-world Prophet'.

Instead, De Cordier has found a tentative yet positive connection with the internet generation, and a common ground with socially engaged artists such as Joseph Beuys.

"There is a whole new generation I can reach out to without having to go through an agent or galleries to earn a commission or a clientele; I can talk to them directly on the internet. This is a modern generation less concerned about labels and categories, asking if I am a designer, an architect, an archaeologist. They don't care. It all ties in neatly with what Beuys called 'social sculpture' and his famous phrase, 'everyone is an artist', a saying which is as simple as it is complex. Beuys didn't suggest that everyone can paint or draw, but everyone can generate ideas, can contribute to society in a creative way."

It is tempting to simplify the attraction bees have for De Cordier. There is the functional, yet very complex and mathematical shape of the hive; the methodical construction of the honeycomb, all hexagons and Roman dodecahedrons. There is also the alchemy of turning flower nectar into honey, considered at one time the equivalent of liquid gold (something which rings true today with the decline of world wide bee population). There is the way bees organise their work and life, their interconnectivity which fascinates De Cordier as much as it inspired Joseph Beuys.

Bees appear in Beuys' work throughout his life. Whilst at art school he produced a series of drawings called 'Queen Bees' which it was suggested was probably inspired by reading the philosopher Rudolf Steiner's 1923 lecture on bees in which Steiner compared the functioning of a beehive to human society. Beuys viewed bees as a symbol of socialism because of the way they live and work together. Beuys was

Lavender, thyme and rosemary are a natural habitat for bees in Alpujarras · Spain

The entrance to the Bibliotaca del Sol in Andalusia, a cultural ark with a small network of mini libraries | 33

34 | Almond trees in the mountains of Alpujarras near the Sierra Nevada - Spain

fascinated by the production of honey and in 1977 he created the piece Honey Pump at the Work Place, where honey literally flows through the building.

"Bees are like the canary in the coalmine," De Cordier says. "When the bees are in trouble, the world is in trouble, they play such a key part in the ecosystem." He first became interested in them after moving to Cádiar, a small town in Granada, Spain. "There was no escaping the bees there. I live in an area called The Greenhouse of Spain. People from everywhere brought their hives so the bees could feed and pollinate."

De Cordier adheres to the theory that the bee has always been a close ally of man and that in a way, the history of honey also mimics the history of mankind. "Bees are prehistoric, like dinosaurs. They are close to men in the sense that we eat the same things: fruits, nuts, vegetables, flowers... For that reason bees can see colours, just as we can. They organise their society in a similar way too; they have a queen, working bees, drones... And it's all about collaborating, not complaining and just getting on with the job. They communicate with each other, through dancing, and can tap into the geomagnetic field of the planet. They are quite amazing."

De Cordier takes the analogy one step further. He argues that men, like bees, try to control nature and manage to create their own little paradise. Men and bees are both gardeners in their own Eden. "It's all about building a sustainable way of life," he enthuses.

For his living sculpture 'House of Bees' (2016, Raversyde), De Cordier built a monumental hive that housed 60,000 honeybees. A bastion, a fortress made of sandbags and jute, materials symbolically linked to conflict and disaster.

We've outstayed our welcome in the garden. Time to leave it to the bees. ✖

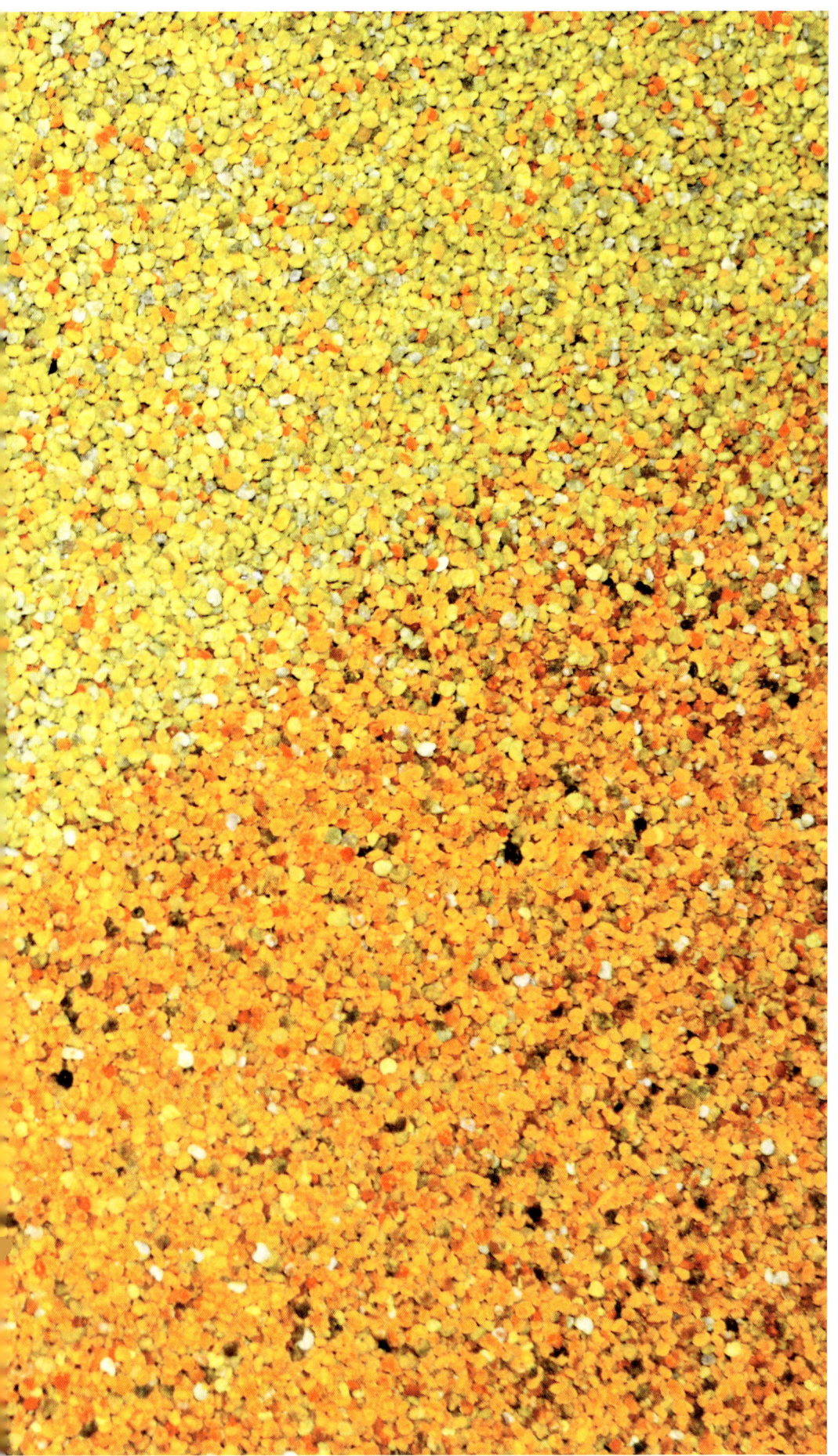

For sessile plants, having sex can be a serious challenge. For (cross-) pollination to occur, the minute pollen has to travel over a large distance to reach the small female flower part on another plant. To make this happen, some plants produce clouds of pollen that randomly travel by wind. When lucky, a single grain may hit the intended target. The rest of the millions of pollen grains go to waste (or even worse, give us hay fever)

As evolution is always on the look-out for smarter solutions, some plants came up with a less wasteful strategy by employing animals for pollen transport. To attract pollinators to the flowers, plants evolved a cornucopia of rewards, ranging from perfume to heat, even the promise of pleasure by flowers mimicking the smell and looks of insect mates. But most plants pay their pollinators an honest wage for their pollination services, by providing floral nectar, the original sugary energy drink.

Almost all animals have a sweet tooth, and humans are no exception. However, in the days before we were introduced to sugar cane, we lacked the ability to harvest sugar directly. Instead, we relied on insects to do the hard work for us. Social bees, in particular, caught our attention, as they ceaselessly labour to create honey reserves. Our craving for the sugary treat was so overpowering that humans were willing to endure painful stings to get to the coveted honey. Similar to our cousin primates who use their hands or sticks to rob the bees' honey bank. Mesolithic cave paintings from Bicorp in Spain show a man harvesting honey from a wild bee colony. Still today, such hazardous collecting of honey is practiced in several countries.

At some point in human history, we must have started looking for less precarious and more steady ways of supplying ourselves with bee products. Rather than pillaging wild colonies, we started to keep bees and look after them. The earliest historical records of this 'beekeeping' go back to ancient Egypt, with later records - and terra cotta bee hives - from ancient Greece and Rome. Also in the new world, Maya records show that they started keeping (stingless) bees almost 1000 years ago. As such, bees are among the earliest domesticated animals.

While humans have been keeping bees for millennia, their role as pollinators eluded us for most of this time. It was only as recently as the 18th century that naturalists Camerarius, Kolreuter and Sprengel discovered the sexuality of (flowering) plants and the role of bees and other insects in pollination.

These insights further raised our appreciation of bees. Not only do they provide us with an assortment of tasty, useful and healthy products, they also fill our shopping carts with fruits, nuts and vegetables. 85% of crops would suffer production losses in absence of pollinators and wild flowering plants would be severely affected as well. The annual economic value of pollination is estimated at $577 billion worldwide. This economic benefit to growers and consumers far outweighs the value of the bee-produced honey, beeswax, royal jelly, pollen and propolis combined. It is for this reason that in recent years, pollination services rightfully have taken center stage.

When focusing on pollination, honeybees are no longer the only candidate. Other insects also pollinate plants and are often even better suited to the job. A particularly effective pollen transporter is the bumblebee. This furry cousin of the honeybee brings a number of traits to the table that makes it especially fit to pollinate. The secret of the bumblebee lies in the fact that this bee is actually warm-blooded. Bumblebees are quite unique among insects as they can raise their body temperature by fast muscle contractions. Combined with their furry coat keeping the heat in, the bumble bee can fly under conditions that are far too cold for other insects. Bumblebees have also other cards up their sleeves. A particular trump is the fact that they use short bouts of buzzing to release pollen from flowers. This is not only highly effective in terms of collecting pollen, but also in terms of pollination.

With all these qualifications in their CV, hiring bumblebees for commercial pollination would seem the evident thing to do. For long, the only catch was that bumblebee colonies could not be produced. Many people tried and failed, until the Belgian veterinarian Ronald de Jonghe discovered the secret to the rearing of bumblebees. In 1987 he said farewell to the cows and pigs and founded the pioneering company Biobest. This allowed him to fully focus on what had been his childhood hobby: bumblebees. As a result, 6,000 years after people started keeping honeybees, farmers can now also keep bumblebees for crop pollination.

Soon, a further benefit came to light: it turned out that bumblebees could even be used to pollinate crops in glasshouses, where honeybees get confused by the diffraction of the light. Tomato growers were all too happy to be given a natural alternative to the tedious and labour-intensive hand-pollinating of every single tomato flower. Soon growers were lining up to buy the hard-working bumblebees and this has been the start of an exceptional and truly green industry.

As a final twist to this story, bumblebees have also been an important motivator for farmers to retire toxic chemicals. When using bumblebees, growers can no longer apply many of the pesticides, as these would kill off, or otherwise harm their much appreciated pollination providers. With the rapid spread of the use of bumblebees in agriculture came an equal boost to biological solutions for crop protection. The humble bumblebee has thus helped make both our food and the environment healthier. ✖

Bees compact the pollen dust into small clumps. Beekeepers collect part of these 'pollen packages', so that we can enjoy them as health food supplements

MASTER OF INSECTS

Jan Fabre (1958) is a Belgian multidisciplinary artist, playwright, stage director, choreographer and designer. He is possibly most famous of all for his Bic Art, ballpoint drawings, which often extend far beyond a piece of paper. In 1990 he covered an entire building in ballpoint drawings. This book shows Bic-art self-portraits of the master as a beekeeper.

JAN FABRE

Hoeder zelfportret I | Ballpoint on colour photo | 1991 | © Angelos bvba

CARLA AROCHA & STEPHANE SCHRAENEN

THE CREATION OF A WALL THROUGH REPETITIVE SHAPES

As a graphic designer, Iris Rombouts is always looking for new shapes and compositions.

"When I discovered the work of Venezuelan Carla Arocha (b.1961) and Belgian Stéphane Schraenen (b. 1971), I was impressed by the simplicity of repetition in their art. The artists started working together in 2005 and they have links to modernist art movements such as Minimalism and 20th century Op Art. Their work, which can be found in the collections of MoMa in New York, MCA in Chicago, MUHKA Antwerp and the Walker Art Center in Minneapolis, often uses repetitive shapes and forms. Their focus is not on objects as manifest things, but on the perception of things as objects. Their installations are flat and composed of many small particles, but nevertheless they are monumental and emphatic. The overt presence of the formal and conceptual in their art can be deceptive. Everywhere within their work – whether bluntly opaque or deceptively transparent – we encounter a narrative. It may appear simple to constantly repeat a certain shape, but Arocha and Schraenen use this concept in an innovative way, choosing the lines and materials with the utmost care. Therefore, their ceramic sculpture 'Wall' and their 'Landscape Meadow 2015' in double-sided Plexiglas are perfectly placed to feature in this book.

The perfectionism in the execution of their work reminds me of the honeycomb, where the bees work so hard to create a repetitive hexagon. 'Landscape Meadow' is constructed from mirror glass panels and steel, resulting in the analysis of a landscape, brought back to nothing but a repetitive shape. The placement of the art works in space has also been though out. For the 'Wall', ceramic modules are stacked together. The result is an impressive Wall of pastels, It has become my own inspirational honeycomb." ✖

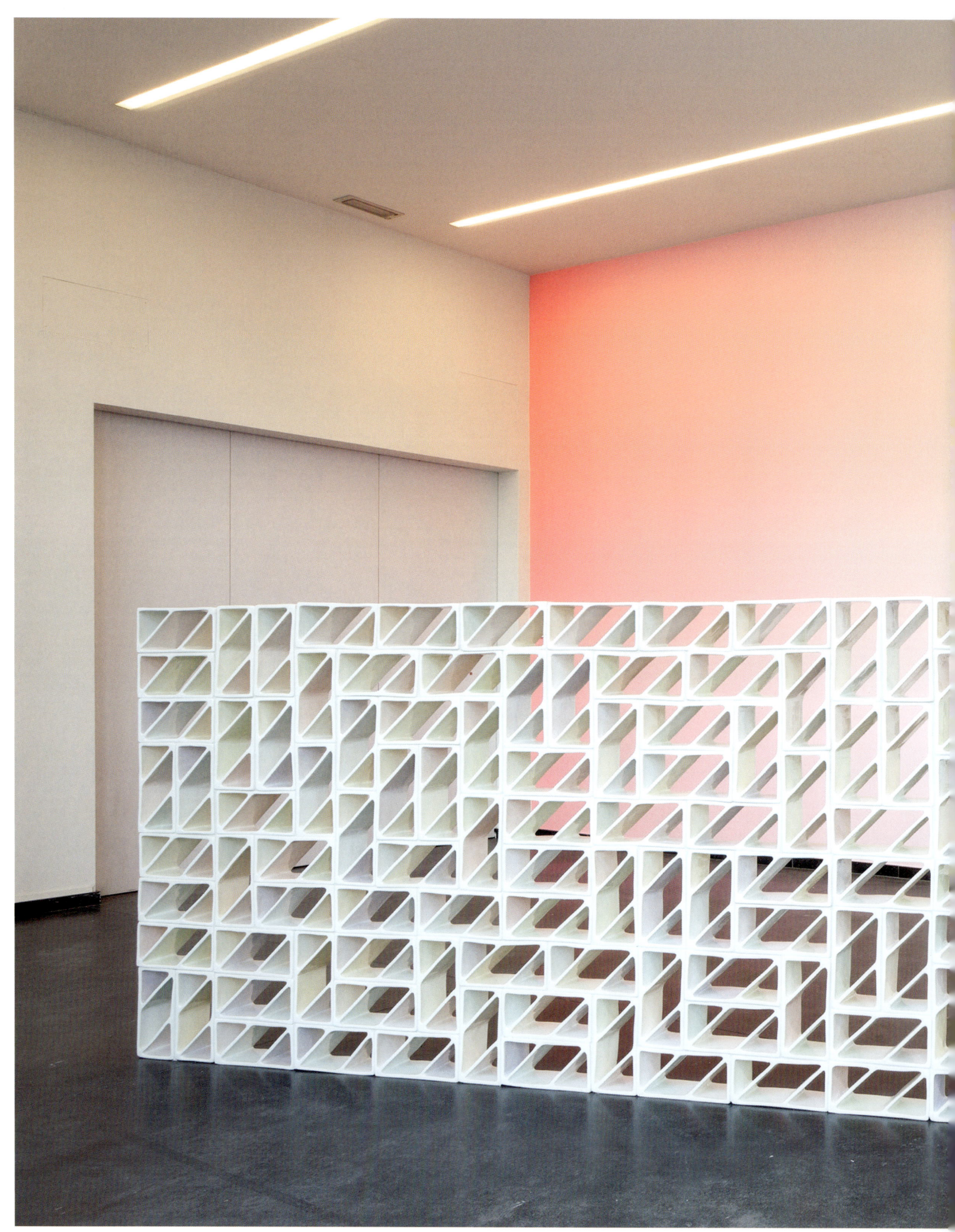

52 | Treshold · an exhibition curated by Carla Arocha & Stéphane Schraenen · Cultureel Centrum Mechelen-Belgium

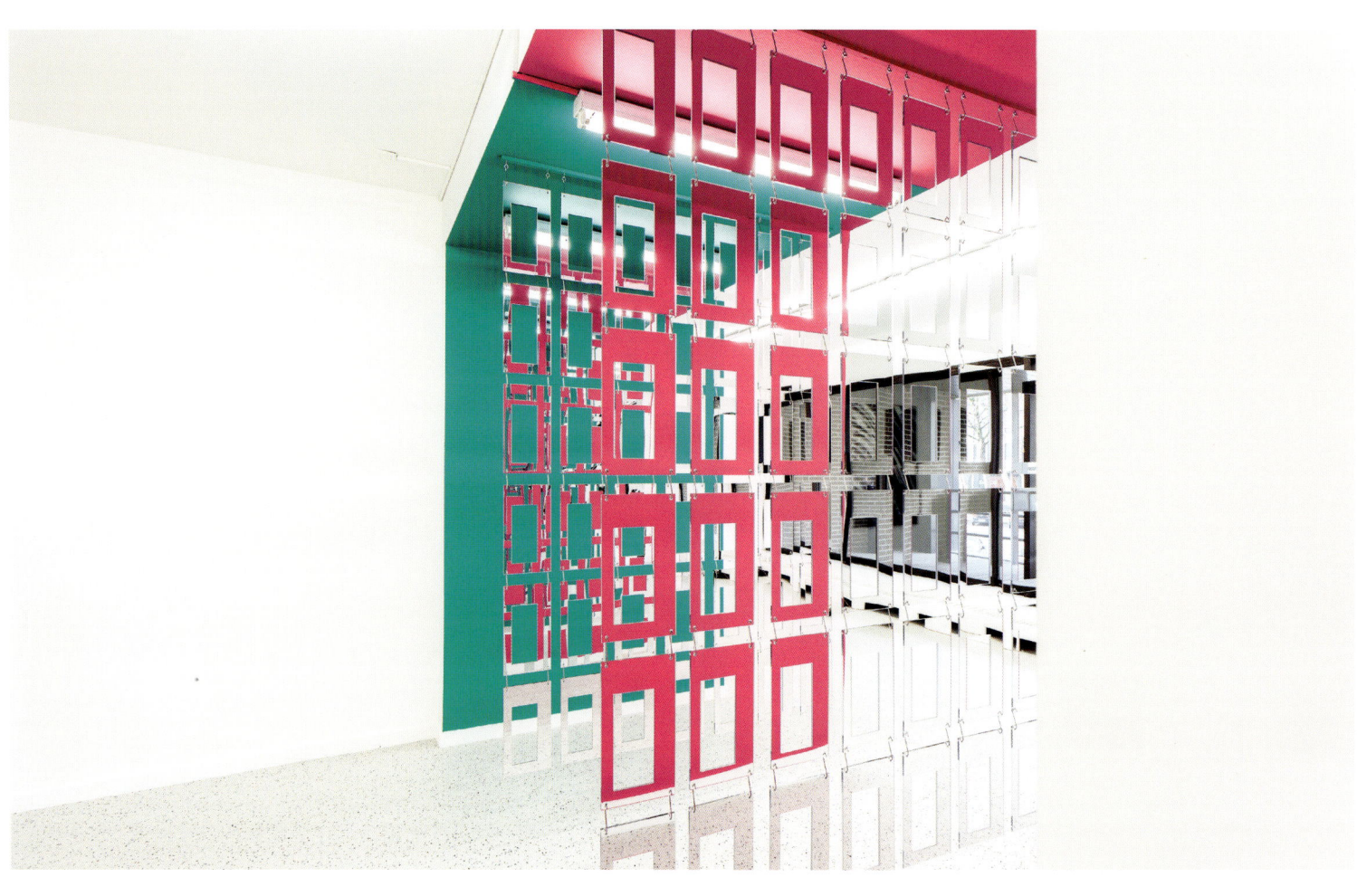

BEEKEEPER FASHION

DREAMER & EXPLORER OF THE ARTS

✖

The Friesian years. Harm van Zwol lives in Antwerp, but judging by the name, you can tell that he wasn't from there originally. And indeed: Harm was born in Friesland, the most northern province in the Netherlands, where people speak another language as well as Dutch. It is an area where nature is unspoilt, people are surly and the clouds hang low. A landscape full of the poetry of nature, reflected in these lines from 'My Kingdom' by Robert Lewis Stevenson:

This was the world and I was king
For me the bees came to sing
For me the swallows flew

We ask him if this is how he feels. "No," he says. "I've had a very protected upbringing, and I was far too fearful to feel like a king." And he adds: "I've always felt very alone."

A thirst for learning. It was his grandmother who gave him his first sewing machine, and though he never even thought of going into fashion, he did make clothes for himself and his friends. This inspired him to choose an artistic path. Friesland was too small for him, Amsterdam too big, so Harm attended the Art Academy in Zwolle, a pretty Hanseatic town with a medieval feel. He studied monumental textile, and started to experiment with forms and materials. When he graduated, he felt there were other disciplines he wanted to explore. So he progressed to sculpting and painting and in the end, he stayed at the Academy for five years. Then it was time to move on, and he took a train to Antwerp - further from home and further in his development. He secured a place at the Royal Academy of Fine Arts, where he expanded his knowledge. But still, he wanted to dig deeper. His intense fascination with materials was a constant factor in his artistic life.

Artist and teacher. Harm stayed in Antwerp and got married. He and his wife had two children: a boy and a girl. It was a bit of a reality check - a family does not come cheap. So his wife - the down-to-earth one in the relationship - encouraged him to get his teaching qualifications. And so the artist became an art teacher at an Antwerp high school, and kept following his own dream as well. Two years ago, he exhibited some of his new and older work, which he calls installations. "At that moment, I suddenly became very aware of something I had been thinking about for quite a while. 'Is this it?' I wondered." It was a kind of 'Eureka' moment for him: he decided to study fashion.

Learning from the best. Sint-Niklaas is small town with the largest market square in Belgium. It also houses an Academy offering part-time courses such as sculpting, painting and photography. But the Academy is particularly well known for its fashion department, led by designers who have studied at the famous Antwerp Fashion Academy. They share their knowledge and fascination for fashion with great enthusiasm and drive, which explains the success of the department. The good thing about part-time education is that it only attracts the most passionate of people, who are prepared to spend long evenings and all weekend to work on often extremely demanding tasks. It takes a special drive to have a full-time job and still study fashion on the side. Here, Harm van Zwol, artist and teacher, realised he had found his new destination. He wanted to dress people. "It felt like something more concrete, less possessive than what I was doing up till then." He calls himself a man with 2D vision: he sees people like paintings. Fashion needs another dimension, and he became fascinated with it. "When I think of fashion it leaves me speechless, I don't feel the need to talk, unlike with sculpting." Why? "Fashion runs away with me, fashion sets me free and everything just happens."

Inspired by the bees. For his second-year collection, he chose beekeepers as his theme. He has always been fond of work-wear: he likes the beauty of its functionality, whether it is the blacksmith's leather apron or the butcher's stripy one. The age-old history of beekeeping goes back to the ancient Egyptians, and is now very much in vogue. This intrigued him: it made a good theme for a fashion student. The result can be seen in the photographs on these pages. Harm doesn't want fashion to constrict people; he likes fluid silhouettes with elements of the beekeeper's suit. "I like to refer to the meaning behind the materials and clothes - one of the materials I use is a linen hospital sheet. I don't want to simply copy something, I'm more fascinated by the ritual meaning of clothes." Details such as the hood, the veil and the gloves are style elements that complete the image. The colours he uses immediately remind us of the work of Beuys, shades of honey, grease and wax.

"Fashion has also taught me about gravity," Harm says. "I discovered that materials flow in a certain way. So different from marble or wood or paint."

Asked what he wanted to be when he was a child, he answers: "A human being." And: "I really want to grow very old."

In the meantime, he continues on his amazing journey. The artist who became an explorer of the arts has now discovered the phenomenon that is fashion, and pushes the boundaries of this new country, where the bees came by to sing for him and inspire him. ✖

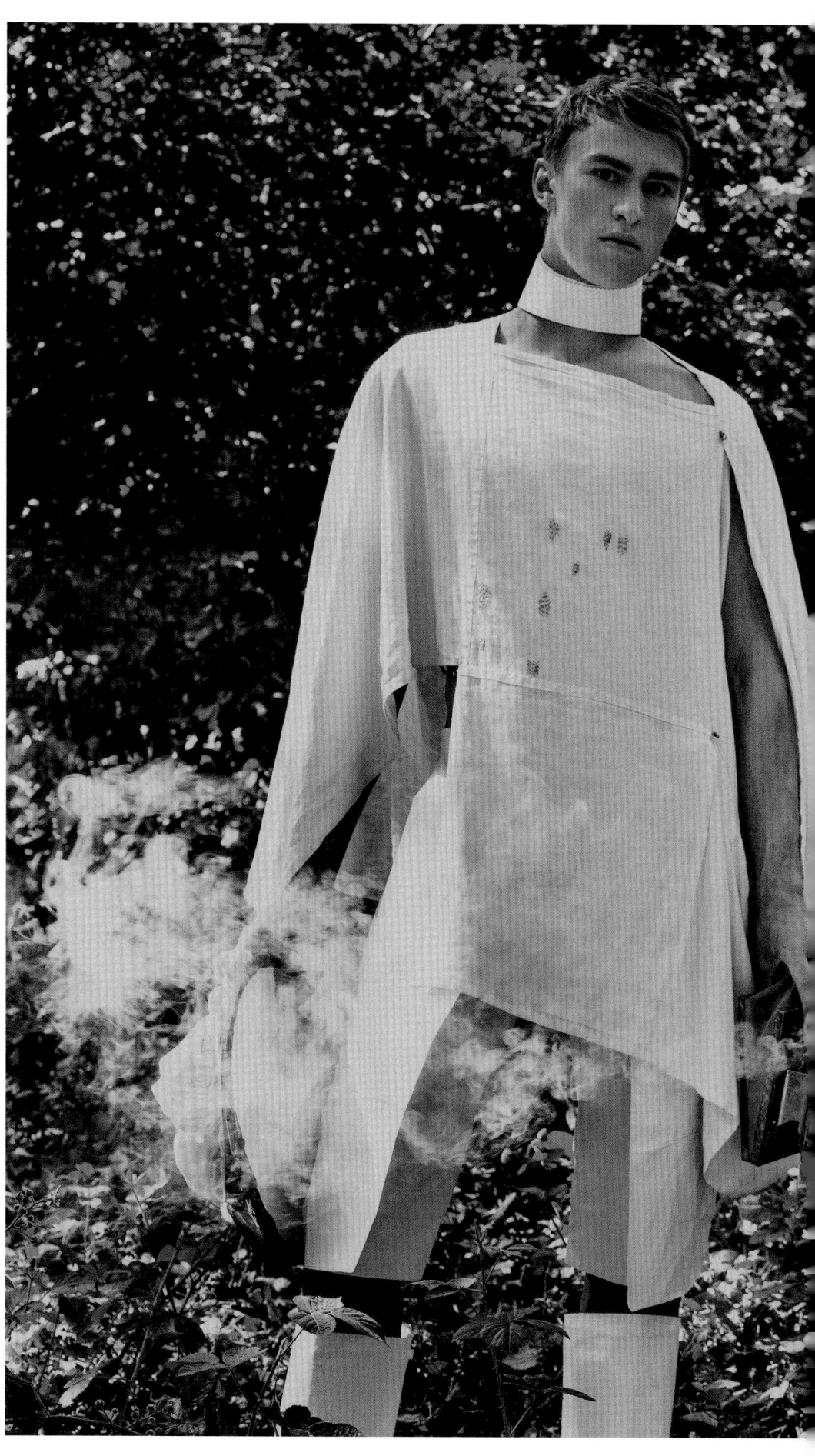

TOMÁŠ LIBERTINY

WAXING LYRICAL
✖

As a student Tomas Libertiny was intrigued by beeswax candles. Over ten years later, he is still fascinated by them: by the wax and by the bee. Recently, he has helped build the world's largest beeswax sculpture in Brussels.

"I was young and a bit of a rebel. I studied art, engineering, sculpture and conceptual design. I was fascinated with literature, philosophy and most of all: metaphysics. What are the underlying principles of things? What is the meaning of objects, materials and reality? That's what intrigued me. One object I was especially interested in at the time was the beeswax candle. It is exceptional from a design point of view. Its shape is completely determined by its the function: the thickness of the wick determines how quickly the candle will burn, as does the thickness of the candle itself. The wick is the skeleton, and the candle burns itself down. It is very simple and very complex at the same time. It is a design that was developed over centuries. It was almost born out of folklore, but it is still an ecological, durable object… At the same time, a candle has something tragic: it disappears when used. A beeswax candle is both ephemeral and durable. A fantastic combination.

I started to look further into the matter. I discovered that beeswax was also used for casting bronze. There, too, the material is of utmost importance, but it disappears into the process. It became clearer and clearer to me that I wanted to work with this material. Not shop-bought, but made by bees. I started talking to beekeepers and watched how they worked. I noticed that they let nature do its own thing, but that they also kept a check on lots of things. Beekeepers are at the service of the bees. By giving them the perfect living conditions they help a bee colony to stay alive and do what it should do: make beehives, make honey and look after the queen. Beekeeping is a wonderful profession. It is an exercise in kindness and patience. To me, as a designer, that is fascinating.

At first, I found beeswax interesting because of its ephemeral aspect, but gradually I started to find the idea of creating together with nature even more intriguing. Industrial design concentrates on perfection. But what is perfect when we kill off the planet with an overdose of plastic, for example? I wanted to incorporate living creatures into my design process, to integrate them into the making of an object. I wanted to sideline myself as a designer, and allow nature to do what it does best. Old-school sculptors used to include wind, humidity and water in the process of sculpting. That way, ageing becomes designing. I like processes that need time, and processes that need a lot of knowledge.

How far could a non-industrial process take me? I saw the bees as a piece of equipment, a cloud of slow nano-3D-printers. I called it Slow Prototyping. I discovered exactly how curved a structure could be for the bees to still want to work on it. Those curves became my alphabet that I used to create shapes that they finished. A kind of scaffolding that I remove afterwards. I knew that one of the first objects I wanted to make would be a vase. Bees get their energy from flowers. The idea that together, we would make an object for flowers was beautiful. The bees worked on a vase for two to ten days, depending on the size of the colony, the season, the weather… Other objects followed, I experimented with pigments, shapes and dimensions. Many of my objects can be seen in museums: in the MoMA in New York, in Museum Boijmans Van Beuningen in Rotterdam, in the Cincinnati Art Museum and in the MUDAC in Lausanne. I make three to four pieces a year, no more. Beeswax is one of the most durable materials there is. It will survive everything, it can last thousands of years. It seems fragile and perishable, but it is extremely strong."

You torture bees, people sometimes say. But bees don't work when they are stressed. They need the right environment in order to grow, live and function. I compare designing with bees to growing a bonsai tree: you have to intervene to stimulate growth and cut where you don't want to see energy wasted.

You cannot compare bees to human beings. Or maybe you can. Bees are very focused on one particular task. They live for six weeks, and for two of those weeks they are constantly building. No holidays, no sleep even. Albert Camus once wrote an essay about the Myth

of Sisyphus. About the sense and the reason of life. About the fact that Sisyphus knows that the work he is doing makes no sense, but he still keeps doing it. Isn't it great to think that we, like the bees, can participate in something we haven't a clue about? Maybe we, too, can't and will never be able to see that what we are actually doing is being determined by something else? To be or not to be, in fact?

I want to continue working with bees. If only to keep them in the spotlight. Bees are vital to our species. Bees create ten of the most important substances known to man, including honey, royal jelly, the bee sting, which has medical uses, propolis and beeswax. Bees purify the air. Birds and bears eat them. They pollinate 85% of the vegetation on earth. And yet we have made the world unliveable for bees.

I have just finished the largest bee sculpture in the world, in Brussels. The Endless Column is 2.3 metres high. It is a wonderful piece. Of course, my greatest dream is to create something on an architectural scale. I am working on it. I want to have the bees make something, then duplicate it in a larger dimension, in various materials. A pavilion? An arch? A façade? Nobody makes shapes as regular and organic as nature itself. No software can even begin to compete. Bees build and calculate at the same time, in symbiosis. They adjust the tension, weight and temperature of the wax while they're building with it. They do it with fluency and flexibility. Unplanned, but entirely according to the laws of their nature." ✖

BEUYS AND THE BEES

SOCIAL SCULPTURE AND INTERNET HIVE MIND

86 | Joseph Beuys | Queen Bee | Watercolour 1958

The buzzword "hive mind" describes social media platforms where Internet users swarm. These virtual interactions determine daily communication, the sharing of information, and ways in which art is approached, created, and viewed. The technology of today presents itself as a tool allowing for German artist Joseph Beuys's decades-old philosophy of 'social sculpture' to be realised on a global scale. The artist's definition of 'social sculpture' stemmed from his belief that everyone has the ability to contribute his or her creative talents to the betterment of society – forming his mantra 'everyone is an artist.' Through Internet access, people are able to widely share their abilities and thus potentially participate in Beuysian social sculpture. Bee-themed art and ideologies in the works of Joseph Beuys compare to contemporary use of the Internet as a hive mind environment for social sculpture.

Beginning in 1947, Beuys pursued his studies in studio practice at the Düsseldorf Academy of Art where he began to articulate his interest in the connection between science and the arts. Beuys writes: 'My starting point in art is more likely through science. Investigating the meaning of life I felt I had a starting point with confrontations with art.' The artist was particularly drawn to the science of bees as metaphor for developing his art practice. For example, Beuys was highly interested the bee's use of chest muscles to produce warmth and secrete wax. The fatty wax produced from this process is used by bees through a fair division of labour in the construction of the honeycomb. The generation of warmth along with the bees' communal production inspired Beuys. He interpreted the honeycomb's functionality (a place for storage and breeding) and the collective effort of creating the hive to be a 'primary sculptural process' of organic and inorganic construction. The warmth and heat produced by the bees created a melting of the wax and therefore made it a more fluid, organic material. Beuys labeled this the chaotic process. The final form of the wax as the honeycomb is a cold, hardened state which Beuys would define as crystallised. The two forms, chaotic and crystallised, would be the foundation of Beuys's sculpture theories. He explains: 'When I speak about thinking I mean it as form. People have to consider ideas as the artist considers sculpture to seek the forms created by thinking. It's the difference between soft, organic forms and hard, crystallised forms: the search is for a solution between these poles. By this I mean to find the evolutionary step towards a new kind of freedom.'

Beuys recognised the polarising forces of organic/chaotic and inorganic/crystallised and surmised sculpture is not solely an art object, but a thought process. In the 1950s, Beuys began to visualise the process through drawings of bees and sculptures incorporating beeswax and animal fat. Queen Bee 3, 1952 is an example of a sculptural work Beuys created while studying at the Düsseldorf Academy. It is created from wax, moulded in the form of a bee, and mounted on wood. The wax is a literal representation of the bee's secretion acting as the malleable, chaotic form. Like the creation of the honeycomb, the wax hardened (crystallised) into its final bee shape. The sculpted bee then serves as simultaneous representation of the organic and inorganic. This wood-mounted wax sculpture was one of the beginning pieces of artistic representation for Beuys's theories on the fluid forces of chaos and crystal, organic life and stability.

By 1960, Joseph Beuys included shamanistic practices in his own artwork in attempts to heal war-torn society as a shaman would heal a wounded spirit. Beuys represented shamanism within artworks by drawing inspiration from the performance format of the interdisciplinary art group, Fluxus – titled in response to the constantly shifting opinions of society and art by nations recovering from the end of World War II. During his involvement with Fluxus (1957-1963), Beuys developed his own performance style, which he called 'actions', relying on (chaotic) gestures, spoken words, and (crystallised) materials such as animal fat and felt to provoke viewer thought. Sometimes the audience would participate in the actions, but more often, they were expected to watch and reflect. After Fluxus, Beuys continued to play the part of the shaman using actions as the primary medium. In 1965, he created the well-known action 'How to Explain Pictures to a Dead Hare' choosing to slather honey (and bits of gold leaf) over his head and face as a sort of ritual anointment. Beuys explains, 'Using honey on my head I am naturally doing something that is concerned with thought. The human capacity is not to give honey, but to think - to give ideas. In this way the deathlike character of thought is made living again. Honey is doubtlessly a living substance.' Honey flows prominently throughout our world. To borrow from V.G. Milum: 'It has been said that the history of honey is the history of mankind.' Milum expounds on honey's tie to ancient use as a commodity, primary sweetening agent, and medium of exchange. Bees produce honey through a natural process for the purpose of survival. It is their instinct to 'gather nectar and lay up stores of honey without limit.' Thus the excess of production of honey allows humans to harvest the substance without harming the bees. Beuys's quote about using honey on his head means humans produce ideas like bees produce honey. For people, ideas are in excess compared to what is needed to survive. Humans are to keep their ideas alive by sharing them as gifts with other humans. In a statement in 1973, Beuys coined his famous phrase "everyone is an artist." The quote is simple, yet complex. Beuys is not suggesting that everyone is a visual artist; rather everyone, through the process of creating and sharing ideas, contributes to society. Beuys explains, "For me the concept of aesthetics in the old sense is no longer relevant. The human being is itself aesthetics, aesthetics today being a side effect of every human activity... If aesthetics equals human being, then human being equals artist."

If every human is an artist by the production and sharing of ideas, then everyone can contribute to the creation of social sculptures, for social sculptures are not necessarily made from marble, wood, or ceramics; they stem from ideas, politics, and pedagogy. Once Beuys began to articulate and act on his utopian ideas of the social sculpture, 'Beuys the artist could barely be distinguished from Beuys the teacher, politician, theoretician, and preacher.' The interdisciplinary ideas of social sculpture influenced the artist's actions. He wished for his idea of social sculpture to become the framework for the 'social organism', or the society at large including military, ecology, economics, and meaning within individual selves. The social organism would be both crystallised and chaotic comparable to the honeycomb and the contributing labour of the bees to the benefit of the hive.

In 1977, Beuys created the piece 'Honey Pump at the Workplace' as a working example of social sculpture. Honey Pump is an action and an installation. Run by electric motors and Plexiglass tubes, the work was designed to literally pump honey throughout the Fridericianum Museum for one hundred days during the Documenta 6 exhibition in Kassel. The piece generated warmth as honey flowed and circulated throughout the building, representing the organic, chaotic component of social sculpture. Sculpted animal fat was positioned on the floor suggesting the shape of a bee around the generators representing the inorganic, crystallised component. The machines were left running throughout the hundred-day exhibit while the action components, 'discussions, seminars, lectures, films, and demonstrations' with the overarching message of art as a subsidisation of life put on by the Free International University completed the social sculpture. Beuys repeatedly stated Honey Pump is only complete when people are present, because the social sculpture needed 'communication, coordination and cooperation to have any meaning when installed,' a Marxist idea where humans are the labour force, producing and sharing ideas. The honey symbolises goods, commodities and economy. Without the humans, the honey does not flow properly. Each person becomes a contributor to the integral system of the discourse not only taking place within the exhibition of 'Honey Pump at the Workplace', but also on a larger scale - the social organism. As stated in Joseph Beuys: Life and Works, 'The honey bee now becomes in its colony-shaping capability a symbol carrier and creates the connection to the social sculpture of society as a work of art.' The social organism is constructed by complex interactions between individuals much as a colony of bees relies on individual workers for the organisation of the whole. 'The question in the early twenty-first century,' Pavlik writes, 'is whether Marx's views still apply in the digital age.'

The colony of honeybees provides a model to understanding interactions between groups of individuals in person and online. Honeybees live and work within a group called a hive. While their efforts as a collective are vital, their individual actions also play a part in the cohesive whole. The hive itself frames the interactions between the bees much as differing groups/organizations in person or on the Internet frame interactions between people. Thomas Seely, a biologist specialising in bees, compared the insects' plight to that of the human stating: 'Living in groups, there's a wisdom to finding a way for members to make better decisions collectively than as individuals.' This system of individuals participating in the group to produce and share ideas for the greater good of the group is often referred to as 'hive mind'. The Web provides endless 'hives' via platforms such as social media sites. The (crystallised) platforms provide a framework for the sharing of ideas (honey) in the process of creating Beuy-

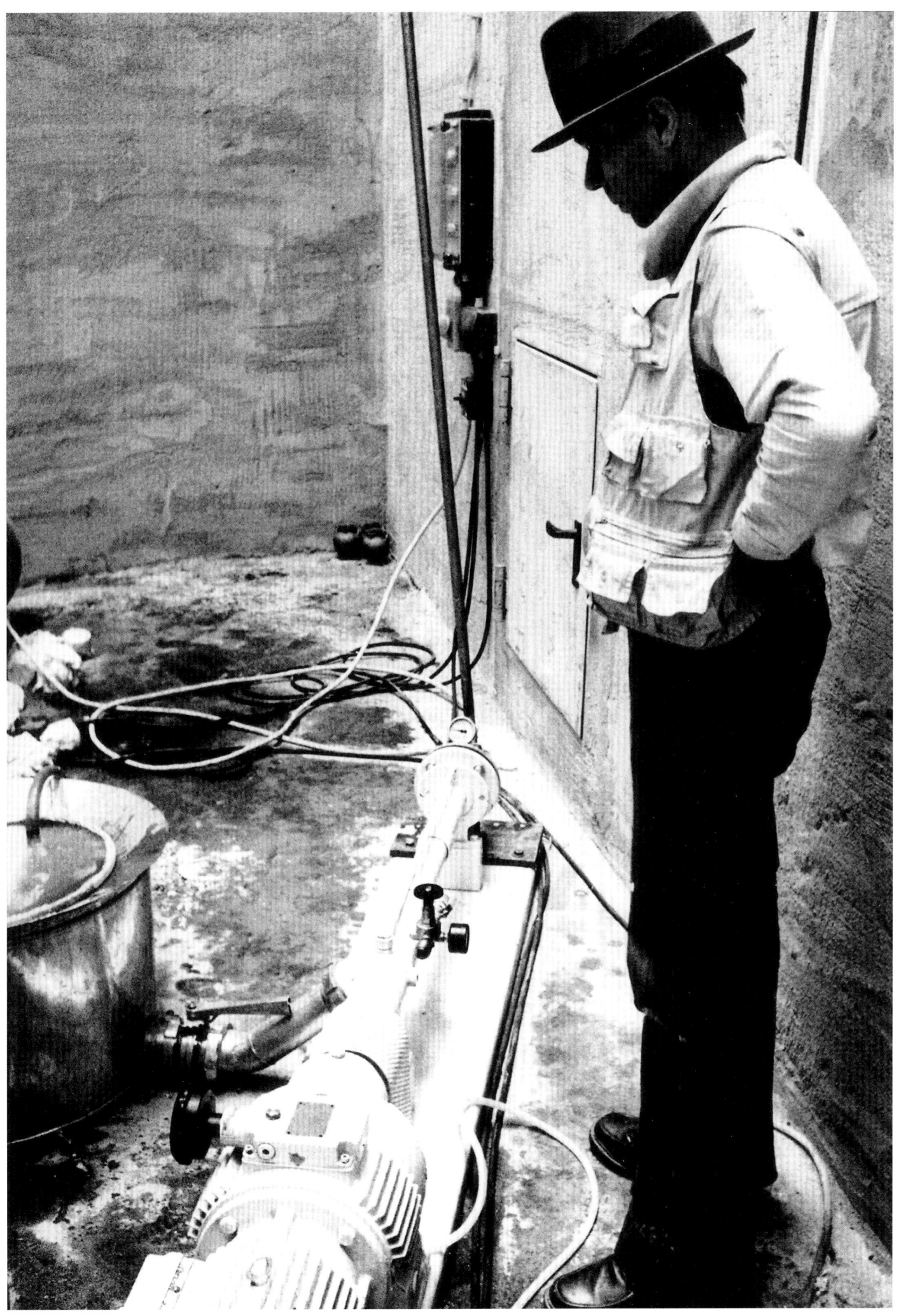

For Beuys the organisational system of bees in the hive was on a par with human social systems | 89

sian social sculpture. As the authors of Connected write: 'Our interactions, fostered and supported by new technologies... create new social phenomena... has significant implications for the collective good.'

Beuys writes about the community of bees as an extension of social and political concepts, 'a socialist organism in which all parts function as a living body.' If these principles are applied in terms of digital technology, then within the social organism that is the Internet there are divisions of groups, like varying hives, in the form of social media, forums, chat rooms, etcetera. Each is a part of the capacious colony of shared information. The information and ideas shared act as the honey in Honey Pump in the Workplace, flowing around the Internet instead of the physical workplace. As long as people are sharing and contributing, the Internet is serving its social function and purpose. Honey is used as a representation for money and commodity, while also promoting the notion that ideas people share are worth just as much. They are gifts and 'everyone is an artist.'

Beuys recognised the inherent human need to connect in careers, academics, and life stating: "All activities, including those of education, training, science, the banks, administration, parliaments, the media, etcetera are integrated into the whole." In the current age of digital networking, the interdisciplinary nature of human connectivity is at the forefront of daily life. The Internet provides the opportunity and the outlet for gatherings (much like Beuys's one hundred days with the Free International University) of shared knowledge, freedom of creativity, interdisciplinary discussion of social issues, and proposing radical solutions. Clay Shirky's 'Here Comes Everybody' gives a variety of examples of people not only using the Internet to learn and share ideas, but also coordinating through social media to create solutions and solve problems. Shirky writes: 'Increased flexibility and power for group action will have more good effects than bad ones, making the current changes on balance, positive.' If the use of the Internet to form groups of individuals who migrate to online platforms to share ideas (make honey) is true, then Joseph Beuys's social sculpture has the potential to be realised through contemporary technology on a global scale.

Andrew Keen's writings and outlook provide a more pessimistic viewpoint on effects of the Internet, social media in particular, changing the way that people interact, produce ideas, and share their knowledge. One of his arguments in 'Digital Vertigo' claims people are drawn to social media sites with a false sense of community. The Internet has made us interact less with each other and more with online identities. This argument, as pointed out by the authors of Connected, is usually debated with each new technological invention. For example, people argued the telephone would change domestic life by losing the 'sanctity' of the home, violate privacy, lead to inappropriate courtship, etcetera. Yet, 'the telephone did more to expand and strengthen local ties than it did to weaken them' and has become 'an organ of the social body,' argues Christakis and Fowler echoing Joseph Bueys's idea of the social organism.
Contemporary business motivational writer Seth Godin's latest work 'Linchpin: Are You Indispensable?' highlights the transition of our current society moving out of the Industrial Age and into the age of the artists. Like Beuys, he claims people have (artistic) abilities, which do not necessarily revolve around creating visual arts, but perfecting and sharing natural gifts and talents. He writes: 'Art is a personal gift that changes the recipient. The medium doesn't matter. The intent does.' The emphasis Godin places on giving and sharing ideas and talents echoes the Beuysian social sculpture. Referencing Internet enabled human connectedness, Godin writes: 'The gift system is now a bigger part of commerce than it has ever been before.' In other words, gifts or ideas or honey flow as valuable currency through our (digital) world economy. Beuys would have undoubtedly used the Internet as a platform for sharing and encouraging others with his ideas on social sculpture as shamanistic/healing gifts to connect humanity. Bees will continue to make honey and humans will continue to produce ideas.

As Joseph Beuys used plexiglass tubing to ensure the flow of honey throughout Fridericianum Museum, so do the miles of fiber-optic tubes carrying Internet access around the globe ensure the continued flow of ideas between colonies. Some groups are forming to assist each other like Beuys's shamanistic healing. Some rally for political causes constructing Beuysian social organisms. Regardless of why groups are formed through the Internet or in person, it is essential to Beuys's idea of social sculpture that individuals contribute their gifts, because "everyone is an artist." ✖

MANCHESTER CITY OF BEES

Floor mosaic in the City Hall

96 | The lead roof of the cathedral became home to the bees in 2012

SOCIETY WITH A SOUL

✖

There are bees all over Manchester. You can't walk two yards without coming across one. In the town hall, a mosaic of bees carpets the floor. Bees feature on dustbins and bollards, trams and manholes and on the coat of arms, which shows a globe and seven bees - for the seven continents, or the seven seas, opinions are divided about that. They are depicted on many of the beautiful old buildings, former warehouses and mills, that now house trendy eateries and shops and tiny, funky apartments. An inked bee climbs up the heel of the girl in the tattoo parlour, one of the many places where free tattoos were given after the bombing at the Ariana Grande concert, in exchange for a donation to help the victims. A mural in the trendy North part of the city features twenty-two bees, one for each person who was killed. The Outhouse, a disused men's toilet - now a canvas for local artists to showcase their work, with paint donated by the art shops in the area - features a giant bee, as a 'message of positivity'. The toilet's roof is covered in wild flowers, making it one of the many green spots in the city that want to attract the real thing. Bees are important in Manchester, more so than ever since the bombing in May 2017. But they have officially been a symbol of the city ever since it began to thrive.

These days, Manchester's greatest export may be the two world-famous football teams, Manchester United and Manchester City. But on the coat of arms, bees have featured since 1842, and make no mistake: they are worker bees. Manchester can trace its origin back to Roman times, when the Romans built a fort on the road from Chester to York, as a settlement for some 500 soldiers. The Romans stayed for 400 years, and the fort was finally abandoned in the 4th century AD. Later, in medieval times, Manchester was a thriving market town with a smallish population. That changed in the 19th century, when Manchester became the world's first industrialised city. Raw cotton was imported from the States, and spun, woven, coloured, printed, packed and sent around the whole world. In the 1870s, Manchester processed two thirds of the world's cotton. The city changed overnight, like a hive that was growing. Factories and mills were built, and packing warehouses. From 70,000 in 1801, the population grew to half a million in 1901 - like bees in the beehive, with everybody working together, doing their bit.

A slow decline started when other countries began to produce cotton goods faster and more cheaply. Between 1870 and 1950 Manchester became poorer, and two World Wars didn't help. But in the last 30 to 40 years a renaissance has taken place. Manchester is now the second most important UK city when it comes to the economy. It has a University population of some 90,000 students, a media city which is constantly expanding since the BBC and ITV moved there, museums, galleries, a teaching hospital, a large IT community and of course, its famous football clubs. The city, once polluted and known for its acid rain, has now recovered environmentally. It is a clean city, with parks everywhere - and beehives.

On the lead roof of Manchester cathedral, Canon Apiarist Adrian Rhodes tends to his beehives, where some 60,000 bees look after their queen and produce sweet-smelling honey. In his beekeeper's suit, hands well protected by plastic gloves - "but I still get stung every now and then" - he blows smoke into the hive to calm the bees, and picks up one of the heavy trays full of honey. His bees swarm out as far as the two rivers of the city, and feed on the flowers in the gardens and the many little parks in the area. They began their cathedral life in 2012, brought here by Adrian: "I had bees at home - I still have, of course. But sometimes, a colony becomes angry, and that's not ideal in a garden with neighbours nearby, so I had to move them. During a staff visit in the cathedral we came up here and I saw this lovely flat roof - ideal for my bees. So I asked permission to bring some hives up, and five weeks later the bees moved here. It coincided with the annual Urban Festival 'Dig the City' to make Manchester a greener place, and the bees have been here ever since. On a clear, warm day, when the bees are buzzing and everything is calm, there is no better place to be."

He lifts a tray, heavy with honey and points out the Queen, who has been marked with a tiny yellow dot that makes her easy to recognise. He shows the pollen, the larvae, and tells how the honey will keep the bees well fed during the winter. Impatiently, he shoos away a male bee that should have long left the hive. "Male bees are useless," he claims. "They just hang around and feed, and they can't even sting."

Downstairs, in his tiny office, he tells more about his work. "Beekeeping is very calming; time goes by and you concentrate on the bees. The cathedral runs a scheme for people who have been without a job for a long time; we have a programme where we help them with CV writing, dressing for interviews, we hold mock interviews and give them a work placement for half a day in the week. Some of them come to help me with the bees, and together we also look after some beehives on another building. It teaches them to do something regularly, to be on time... they learn they can do something they would never have thought of doing. When they go for an interview, they can say: 'I've been beekeeping' and that makes them interesting. I'm a psychotherapist as well, and over the beehives people tell me their troubles, we talk about all sorts of things. We do quite well in getting people back to work."

Why the bee as a symbol of Manchester? He explains: "The bee has always been thought of as a symbol of a perfect society. In the Middle Ages people thought the beehive was run by a king. He was the head and all the bees had their job; there were soldiers, and workers in the field, and some looked after the babies. Just like in a medieval society, everybody had their place. Apart from that, the bee will lay down his life for the beehive, for the king, as they thought. When a bee stings he dies. In medieval times, it was considered that bees represented a perfect society."

Times changed, the bee symbol remained. He continues: "In the 19th century, Manchester became the first industrial city, with its cotton and silk mills, and people flocked there and had to start working in a different way. In the fields and the villages or in their homes, they would work in a group of three of four. Factories had hundreds, thousands of people, so when they became factory workers, it was the first time that people worked together in such large groups. Everybody had his or her job in a factory, and then there was a manager. A factory was like a beehive, with the many at the bottom and the managers at the top. It became a symbol of good community working, even though the capitalists owned the mill. Although Manchester was very much a socialist city in the last part of the 20th century, in the 19th

Like the bees, Manchester people are a close-knit working community

century it was run by the capitalists. To them, the symbol of the bee was very convenient; it said: we will become wealthy if everybody does their job."

After the bombing in the Arena, the bee has become a stronger symbol than ever. "It was about 300 metres from the Cathedral; we couldn't get in for a few days. The whole city was deeply shocked. The fact that so may young people had been killed, struck deep into the heart of the people of Manchester. As in a beehive, we close ranks when we're under attack. We are a stubborn people. This is not the first time that there's been a bomb; there were two attacks by the IRA in the 90s. We are kind, neighbourly and friendly, but we are tough. People came together to grieve, as if to say: we are a strong community, we live together, we defend ourselves. You will not defeat us, you have picked the wrong city. If you poke a stick in a beehive, the bees will come and sting you. That's how we are. The Arena will reopen soon. We don't want revenge, but we are determined not to be beaten, not to let evil overcome us. We have a large, very mixed community in Manchester, with Muslims, Chinese, Jewish, people from everywhere, and we will not let this divide us."

The cathedral, now also the dwelling of the bees, started out as a small church in the Middle Ages. In 1421 it became a Collegiate Church, with a college of priests and singing men, under a warden. A complete rebuilding of the church was undertaken, which took about 100 years. In 1847 the church became Manchester Cathedral, an imposing building with a wide nave and beautifully restored stonework. In 1940, it suffered terrible wartime damage, but it is still standing proudly. It is just one of many beautiful historic buildings in this city: warehouses, a corn exchange, the impressive library...

Canon Rhodes explains: "When the cotton industry went abroad in the sixties - India and the United States had cheaper production methods and labour- this part of the world became very poor. Manchester and the surrounding towns depended on cotton. In the sixties, seventies and eighties, the city became very run down and there was no money to rebuild . Which means that nothing much was knocked down and a lot of the old buildings were just left to stand empty. After the IRA bombing in 1996, a lot of money was poured into Manchester. The city became very strong again, determined to get over the devastation. Much of the money was used to turn the old buildings into something beautiful. Many people now live in the old mills; the 150-year old buildings were made into apartments. The Old Corn Exchange, where business used to be done, is now full of places to eat. Where there were empty spaces, wonderful new buildings were put up. We have the best of the old and the best of the new, unlike other places. Take Birmingham: it was very rich in the sixties and everything was ripped out and replaced with horrible concrete buildings. We were very lucky that we became poor for a while..."

In the Northern Quarter, where the trendy people hang out, funky coffee shops and mouth-watering burger restaurants stand shoulder to shoulder with modern glass buildings. Alleyways with old fire escapes create a New York-

Urban beekeeping is becoming more and more popular

104 | Skyline view from the Hilton rooftop bar

106 | Film makers often use the atmospheric North of the city for scenes set in America

like atmosphere that is reminiscent of West Side Story. Filmmakers often use this area for scenes set in America. Afflecks, once a four-storey department store, now houses some 80 small shops with a colourful retro indie hippie vibe: tattoo parlours, fancy dress shops, cafés and a staircase that is, once again, an ode to the bee. A bit further, a huge new mural depicts 22 bees, one for each of the people who died in the Arena.

Outside the cathedral, people sip coffee in sidewalk cafes and shop in Harvey Nichols and Selfridges. At Cloud 23 in the Hilton Hotel, afternoon tea is served, with stacks of delicate sandwiches and tasty scones, set against a background of the beautiful skyline. Waiters come and go, businessmen make important phone calls and well-dressed friends open their shopping bags and admire each others' latest purchases. In the magnificent Edwardian baroque Midland Hotel, the table in the Michelin-starred restaurant is being laid for dinner. It was here that Mr. Royce and Mr. Rolls first met, which led to the formation of Rolls Royce Limited in 1904. Later that evening, young go-getters at Albert's Schloss restaurant press the 'push for prosecco' button on their table and drink to health, wealth and happiness. And around the corner, the homeless people of Manchester settle down in their sleeping bags, echoing the male bees that have been pushed out of the hive. The new mayor has made them his highest priority, and a scheme is being put into place to give them a home.

A bee-shaped dustcart rides by, a tram lights up. On the bee wall at Afflecks, a black and white etching of a bee claims that 'Manchester is still buzzing.' It is. ✖

#WE ARE MANCHESTER.. #WE STAND TOGETHER..

NO Fear
HERE

BEAUTIFUL HONEY BODY

It's been known for centuries: honey does wonders for your skin. But it's only in the last few years that scientists have begun studying its benefits.

When you taste a drop of honey it is hard to imagine that the production process of this liquid gold is rather unsavoury. Bees scoop the nectar from the flowers with their tongues. They store it in a special stomach, where enzymes break down the complex sugars into a sugar that crystallises less quickly and can be kept longer. Bees can transport their own body weight in nectar. In the hive, they throw up the nectar and it is passed from mouth to mouth to the other bees. During this digestion process, most of the liquid evaporates, making the nectar much more solid. The bees fill the cells of the honeycomb with it and flutter their wings to allow as much liquid as possible to evaporate. When less than 20% of the liquid remains, they close off the honeycomb with wax, allowing the enzymes to complete the transformation process from nectar to honey.

The wax is also produced by the bees; they 'sweat out' the sugars in the nectar. By chewing on it, they turn it into a material they use to build the cells in the beehive. Beeswax is not only a good base for creams, it also makes the skin supple and soft. Another 'building material' for the beehive is propolis. The bees get this mix of pollen and resins from plants and trees. They process it into a balm that they use to plaster the inside of the hive. It keeps the hive dry and free from bacteria, protecting the colony against infections. It is like a vegetable antibiotic. People use it in food supplements as a fortifying concentrate. But the most exclusive substance a beehive produces, is royal jelly, the superfood for the larvae that have been chosen to become queen. It is extremely nutritional, causing the future queen bee to grow and gain weight in record time. Compared to honey, it is produced in minuscule quantities, making it extremely

Egyptian queen Nefertiti and Poppea, the wife of the cruel Roman emperor Nero, used honey in their facial toners.

precious. That is why it is often made synthetically to be used in food supplements or anti-ageing creams.

However, it is honey that is used most often for skincare. The most famous beauties in ancient times used it in their routine: Egyptian queen Nefertiti and Poppea, the wife of the cruel Roman emperor Nero, used honey in their facial toners. Cleopatra, too, poured vast quantities of honey into the asses' milk baths she took to keep her skin soft. The ladies-in-waiting in the Chinese Ming dynasty mixed honey with ground orange pips for a clear complexion.

"Honey consists of 80% sugar," says Marine André of Bee Nature, a Belgian skin care brand based on honey. "Sugar retains liquid very well, which makes it ideal for a dry or fragile skin. It is also good for mixed or greasy skins, because it limits sebum production and acts as a disinfectant, giving bacteria no chance. It also has a calming effect, reducing inflammation and it can be tolerated by even the most sensitive skins." But there are other ingredients – amino acids, trace elements, minerals and vitamins B and C – that also make honey interesting, says Marine. "It provides all the nutrients the skin needs. Moreover, the antioxidants (flavonoids, polyphenols, vitamins B and C) neutralise the radicals, slowing down the ageing process in the skin." An ideal skin care product, you could say. Yet, Marine noticed that the large cosmetics firms didn't really pay much attention to honey, even though pharmacists often use it in their wonderful concoctions. "Six years ago, honey was hardly ever mentioned." Her brand, Bee Nature, was the first natural brand in pharmacies. Since then, more and more attention is being paid to natural ingredients, and the success of DIY cosmetics has greatly increased. Honey, therefore, is no longer limited to the list of ingredients of obscure biological brands; the well-known brands that are sold in supermarkets and perfumeries have also discovered this liquid gold.

Not only beauties, but also doctors from the Ancient World knew about the many merits of honey. The oldest mention is in the Ebers papyrus – one of the oldest medical documents from the 18th dynasty that started in 1550 B.C. It mentions 147 recipes with honey, to cure baldness, but also abscesses, sores and wounds. The ancient Egyptians, Romans and Greeks all knew about the disinfectant and antibacterial properties of honey, which they used in creams to treat burns, cuts, grazes and other wounds. The natural healers in Europe, Africa, America and India did the same. Ayurvedic natural medicine has no less than 634 remedies containing honey. They are used to treat as many ailments, from encouraging hair growth to getting rid of tapeworms. In traditional Chinese medicine, honey is still used to relax, detoxify and as pain relief.

With its unctuous texture, honey is an ideal basis for ointments that can keep for a considerable amount of time. In Ancient Egypt and Assyria they knew that honey, which is acid rich and contains very few bacteria, is perfect for embalming bodies. Honey ensures that the cadaver doesn't rot, but dissolves, as it were. The Arabs went even further in their experiments, according to the writings of Chinese pharmacist Li Shizen (16th century). He wrote about the lugubrious technique of the 'mellified man': older volunteers ate nothing but honey and even bathed in it, until their sweat, urine and faeces consisted of nothing but pure honey. When their diet finally proved fatal, they were buried in a stone coffin, filled with honey. A century later, these 'mellified men' were dug up and their remains were used in a medicine that would cure all illness. It was said that the substance was sold at an extortionate price, because of its long 'preparation time' and the lack of volunteers.

In spite of these wonderful properties, honey more or less faded into oblivion during the last century. With the discovery of penicillin in 1928, honey was de-

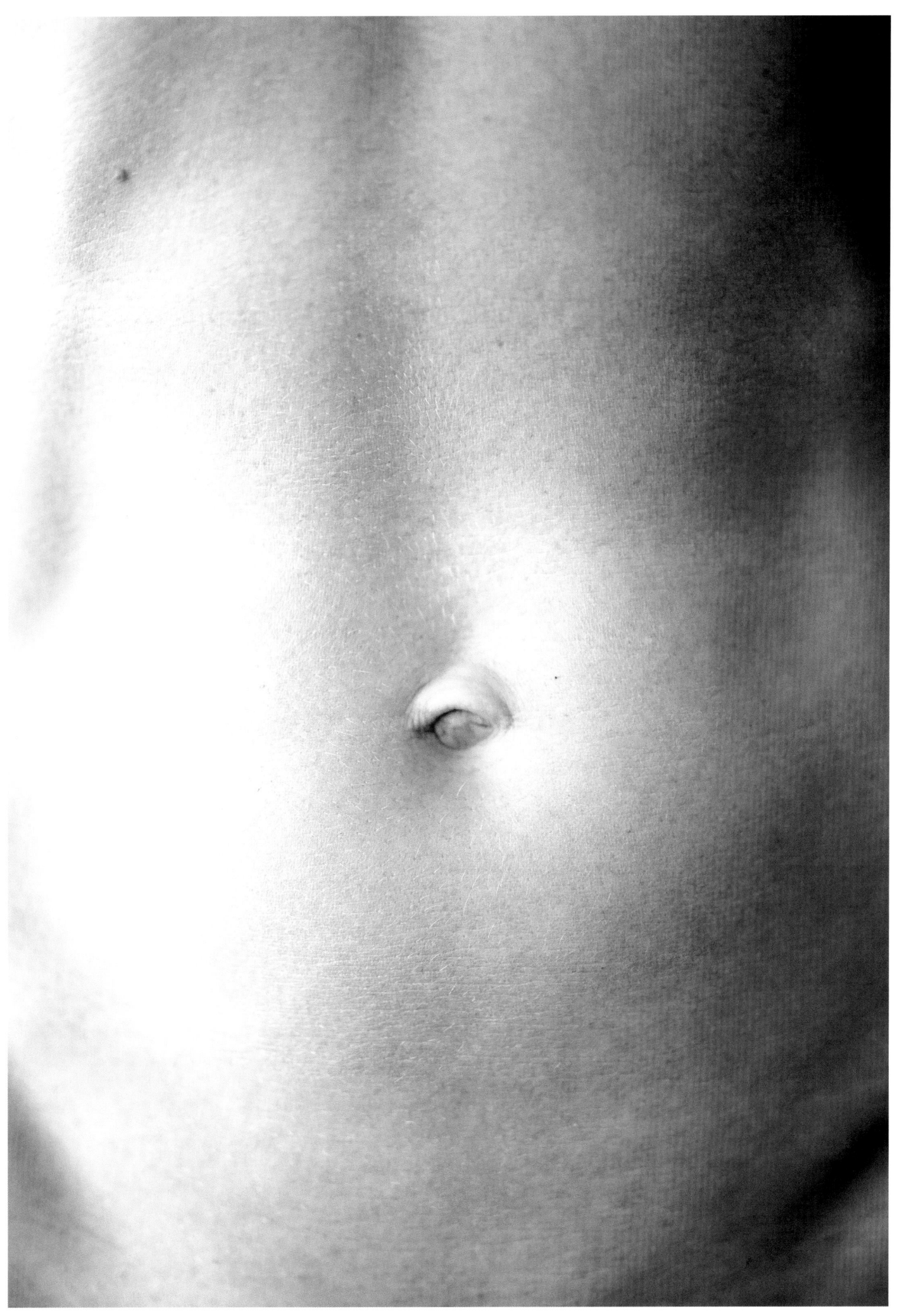

In Ancient Egypt and Assyria honey, which is acid rich and contains few bacteria, was used for embalming | 117

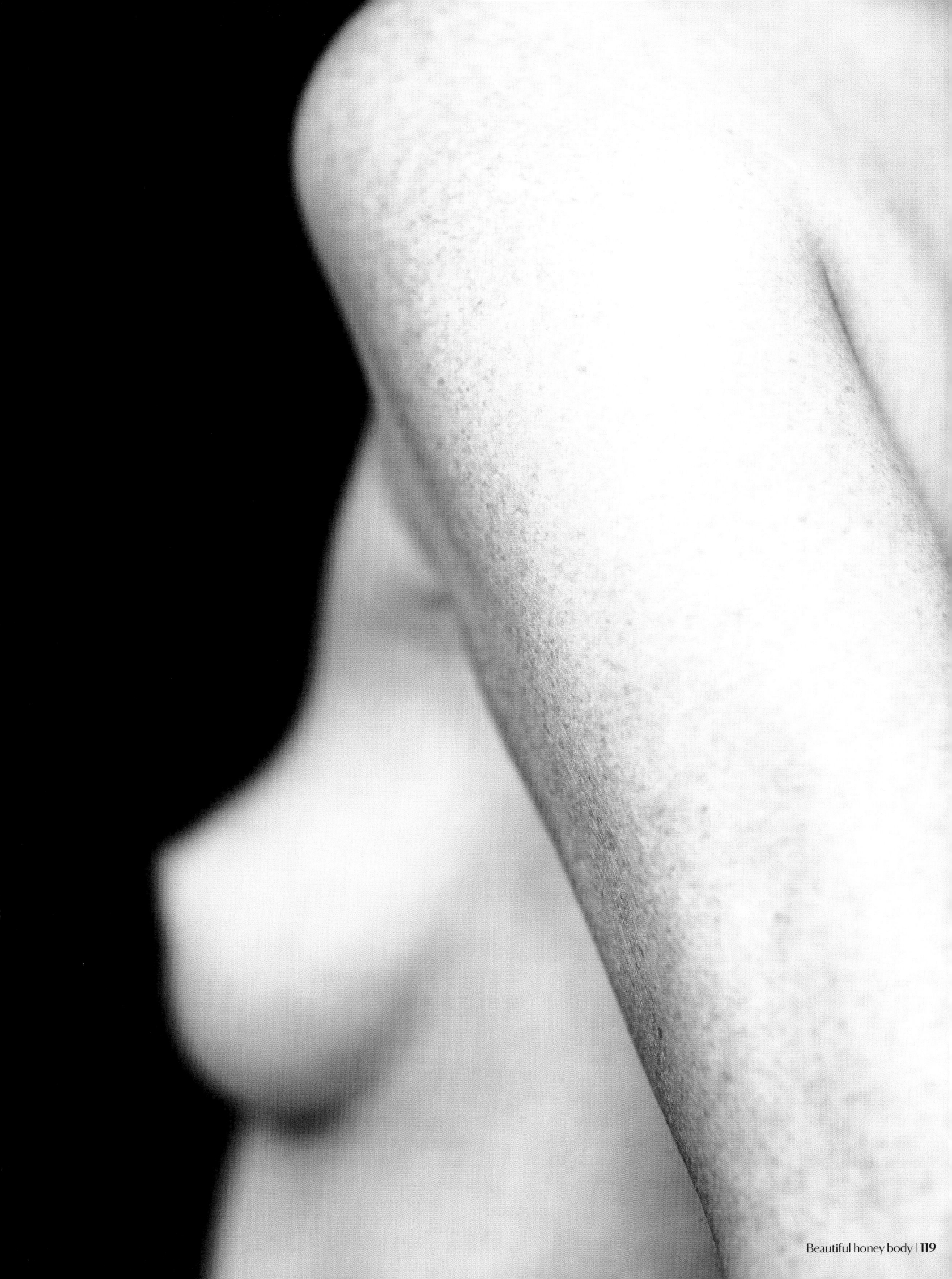

graded to a home remedy, we are told in the BBC programme 'Hive Alive'. It was the first antibiotic that was used on a large scale to fight infections. But nearly 100 years later, it became obvious that viruses mutate at an incredible speed into versions that become more and more resistant to antibiotics. And now that scientists don't know where to go from here, honey once again gets the attention it deserves, says medical author Nathaniel Altman. 'Research has shown that honey has a strong inhibiting effect on the growth of some sixty different bacteria,' he writes in his book The Honey Prescription. 'Bacteria become more and more resistant against antibiotics, but they don't stand a chance against honey. Moreover, honey doesn't have any side effects, as opposed to antibiotics and other medicines. It is cheap and suitable for everyone.' All these points counterbalance the sceptical attitude of medical scientists towards natural products, says British microbiologist Dr Matthew Dryden in the BBC programme. Dryden developed a medical honey that is supposed to be even more efficient than the famous New Zealand Manuka honey. SurgiHoney is now used in various hospitals in the UK and in developing countries such as Ethiopia, Uganda and the Sudan.

One of the first and largest studies on honey was conducted by the late Dr Bernard Descottes in 1984. The French surgeon treated more than 3000 patients with honey, applying it straight onto their wounds. He discovered that thyme honey has the best antiseptic effect and is therefore the most suitable for medical purposes. His example was followed by several colleagues, one of them the French professor Albert Becker. The latter compared various studies and concluded that there are several reasons why honey is an unfavourable breeding ground for bacteria. First of all, bacteria have trouble travelling through the protective membrane the sticky texture leaves on the wound. The acidic environment (PH level between 3.2 and 4.5) makes it hard for bacteria to survive. As does the lack of wound fluids, because these are absorbed by the sugars, allowing the wound to heal faster as well. And then there are the antiseptic substances, the natural antibiotic inhibin and the oxygen peroxide that the honey produces, and that destroy the most stubborn of bacteria for good. Because of the antioxidants and calming substances in the honey, wounds heal faster, better and more smoothly. For five years, French surgeon David Lechaux of St Brieuc hospital has been using honey to treat open wounds and tissue damaged by radiation. He conducted a study on 60 patients, comparing the effect of honey with that of traditional medicines, and the results were spectacular. Because it is also a cheap treatment, he wants to introduce the technique in developing countries. But it is also used more and more in hospitals here, for instance in the Neder Over Heembeek burns centre and the Brussels Brugmann hospital.

Many studies are being carried out outside Europe as well. In New Zealand, the local Manuka honey, considered one of the most healing kinds in the world, is being studied. Professor Peter Molan of the Waikato University found that this honey conquers sore throats, stomach and intestinal infections, caries and fungi. The university even has a special department, the Honey Research Unit, that tests the effect of honey on amputations, abscesses, chapped nipples, sores, fistulae, grazes, cuts and infected wounds. In Japan, Professor Miki Fukuda of Kyoto University published a study about the improved immunity and anti-tumoural activity in mice after treatment with Nigerian jungle honey. Other studies, too, confirm the positive effect of honey on the treatment of certain cancers. The future seems sweet... ✖

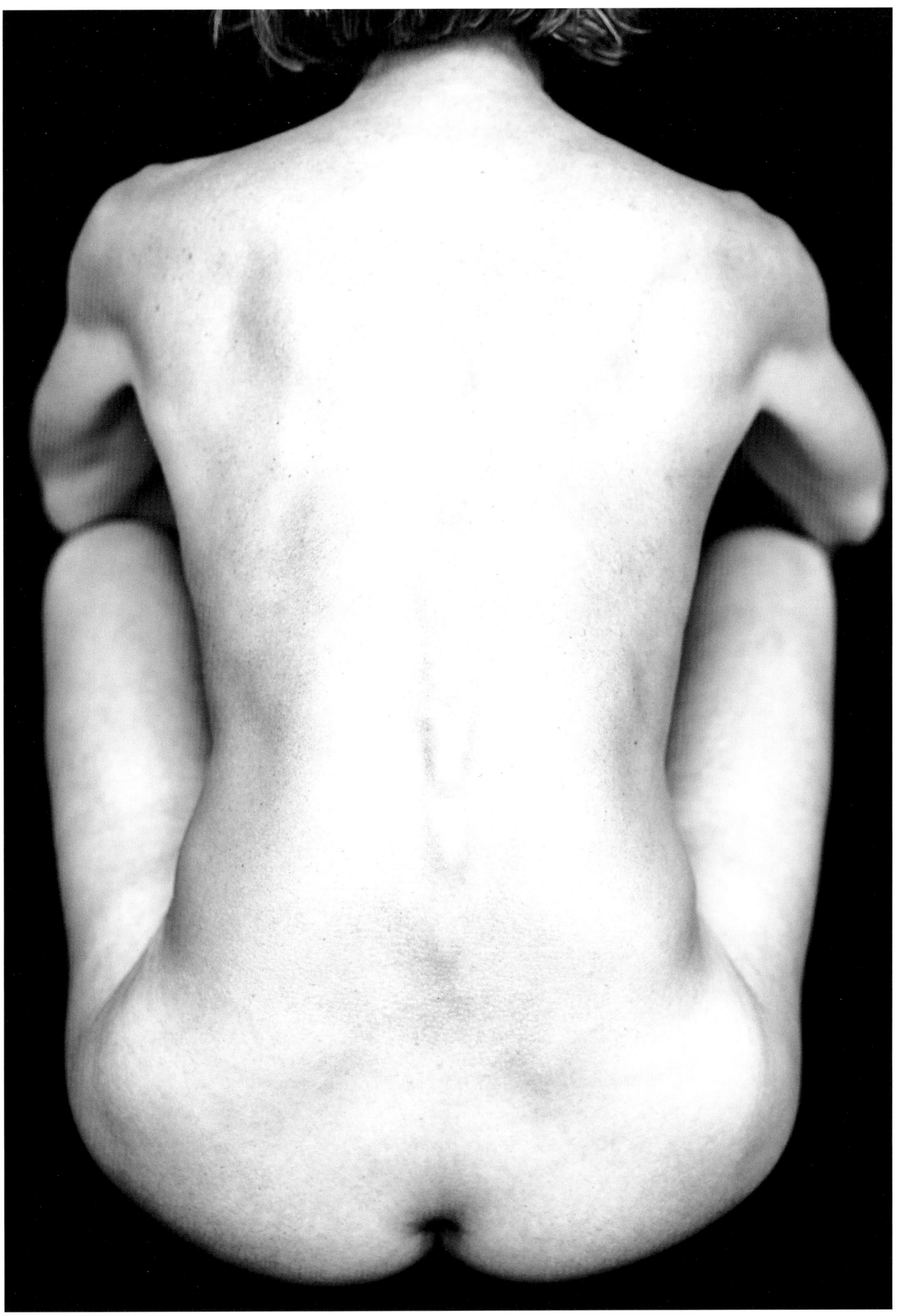

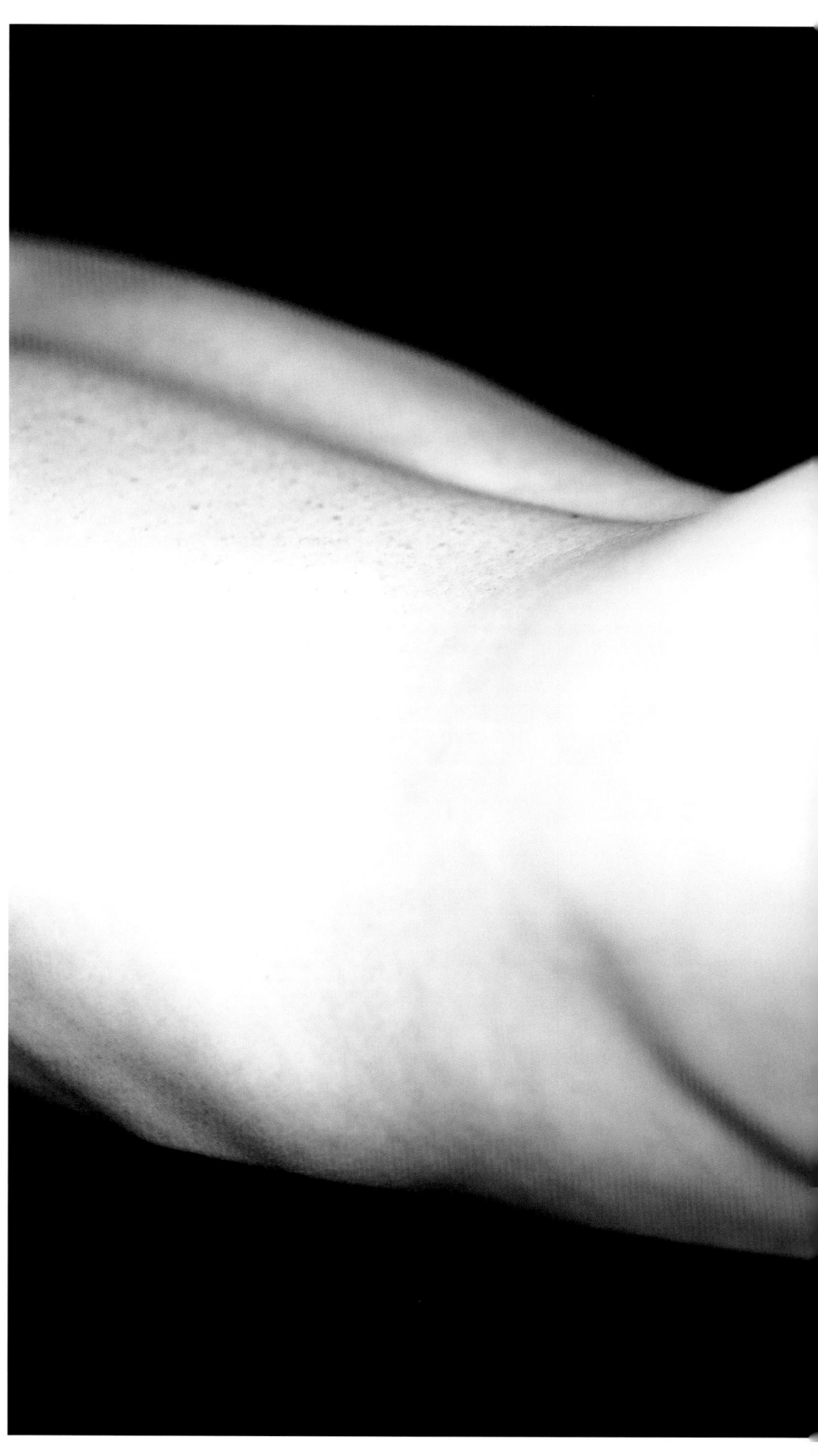

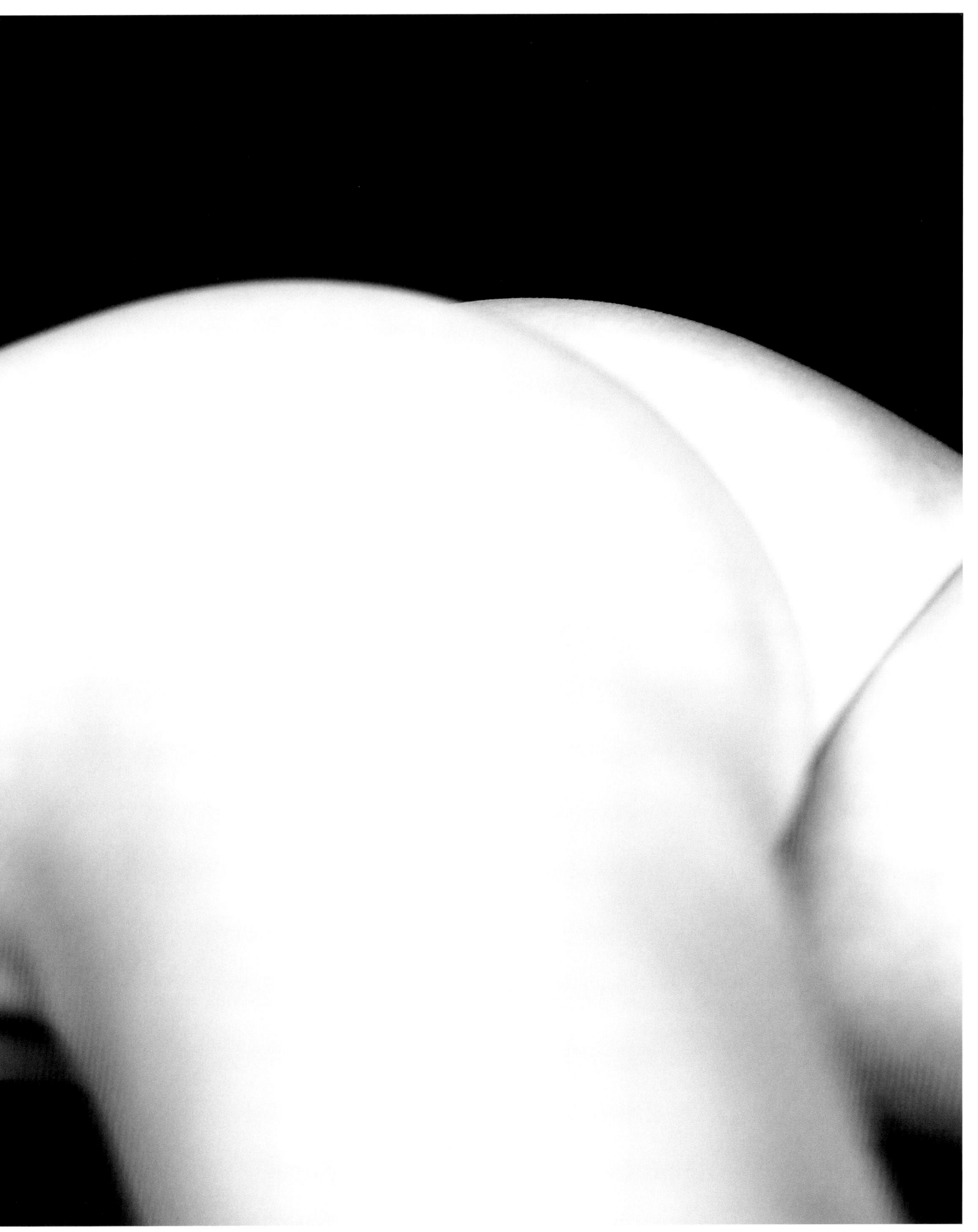

THE COLONY GREW IN MY BODY ALL THAT SUMMER / THE GAPS BETWEEN MY BONES FILLED WITH HONEYCOMB AND MY CHEST VIBRATED AND HUMMED / I KNEW THE BROOD WAS HEALTHY BECAUSE THE PHEROMONES SANG THROUGH THE HIVE AND THE QUEEN LAID A GOOD TWO THOUSAND EGGS A DAY / I SMELLED OF BEE BREAD AND ROYAL JELLY / MY NAILS SHONE WITH PROPOLIS / I SPENT MY DAYS FREEING BEES FROM MY HAIR / AND PLANTING CLOVER AND BEE SAGE AND WOUNDWORT AND TEASEL AND BORAGE / I WAS A QUEENDOM UNTO MYSELF

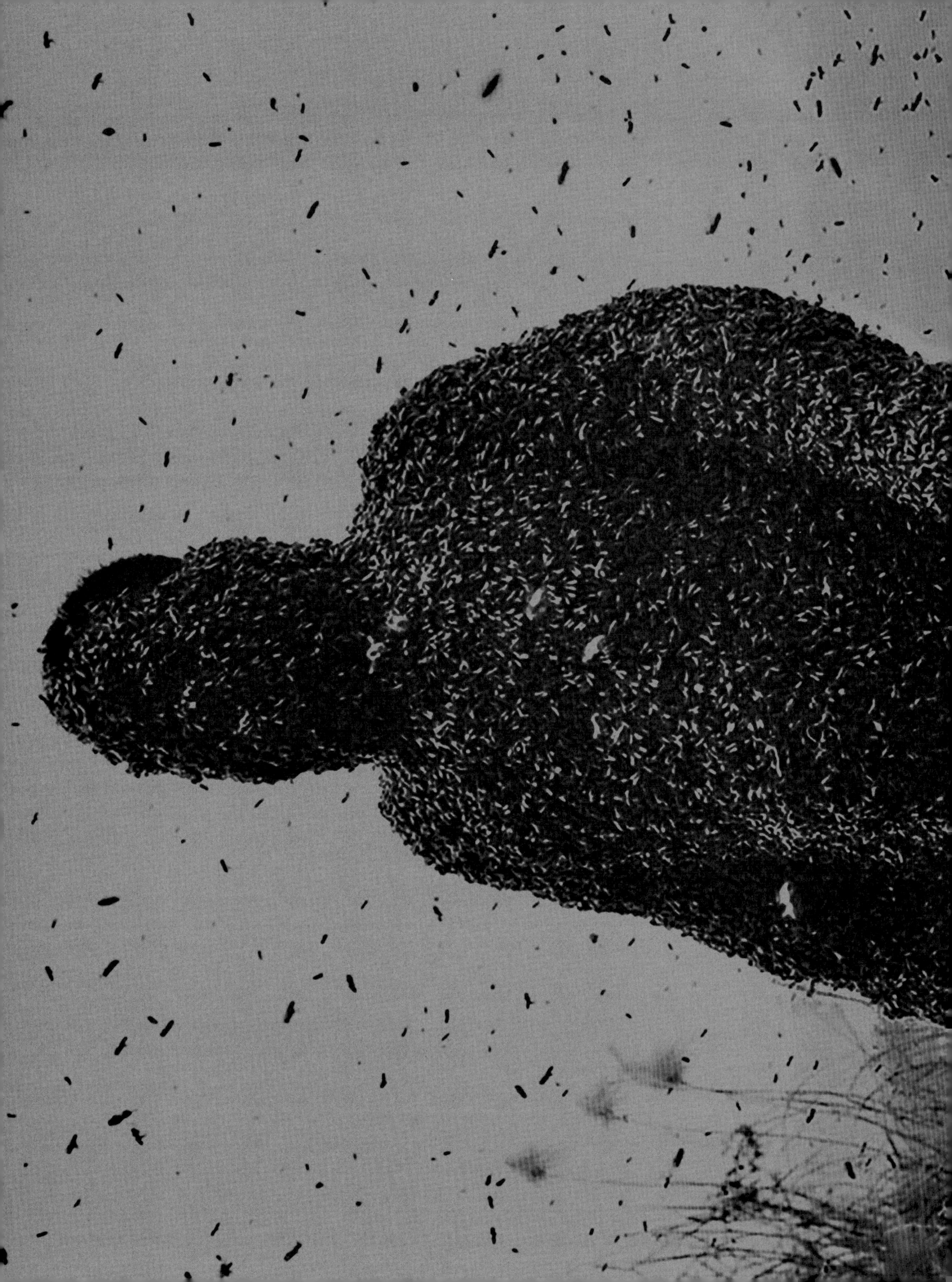

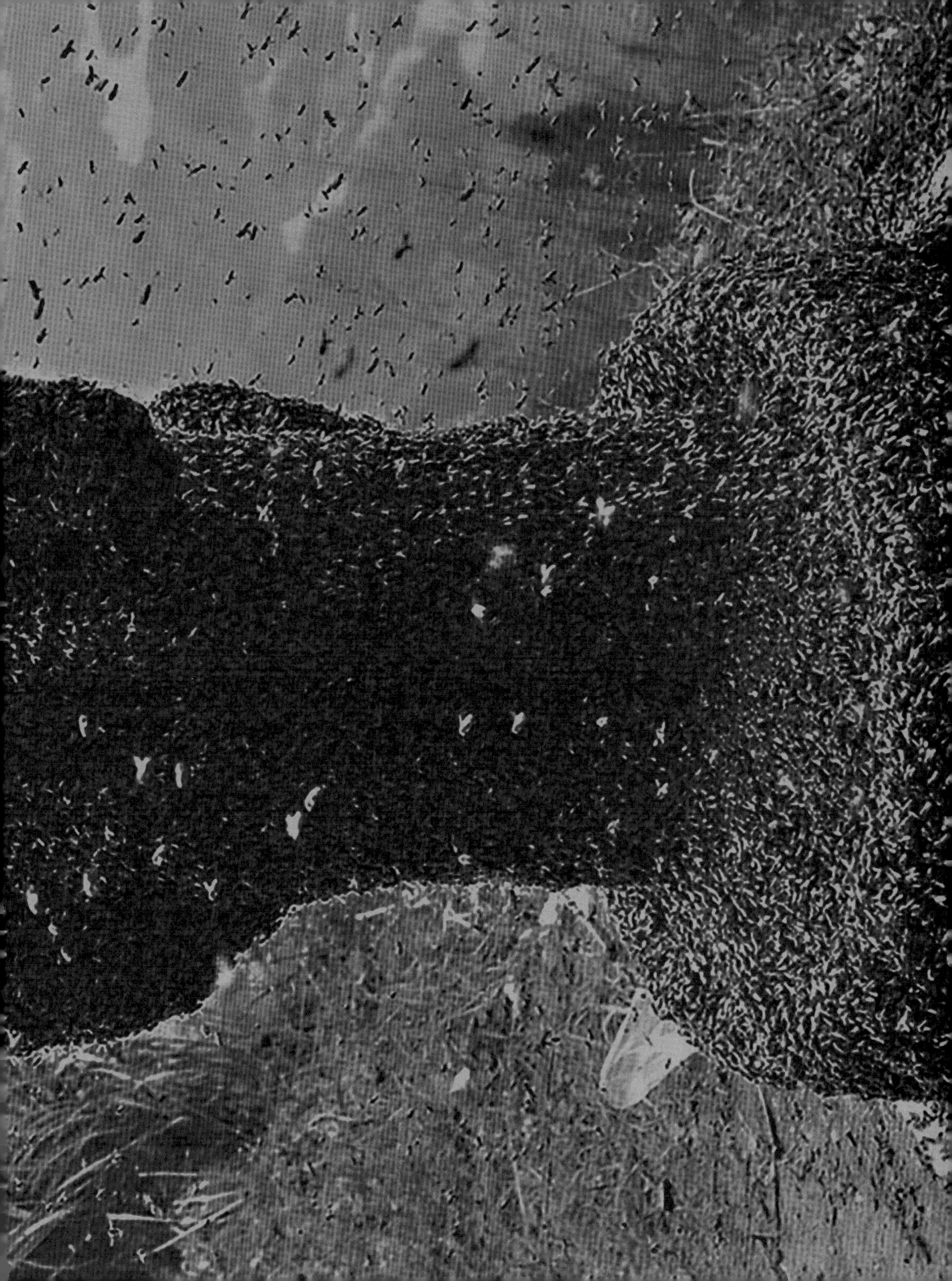

FAME IS A BEE / IT HAS A SONG / IT HAS A STING / AH TOO IT HAS A WING

THE BIOGRAPHY OF THE BEE IS WRITTEN IN HONEY AND IS DRAWING TO A CLOSE / SOON THE BUZZING PLAINCHANT OF SUMMER WILL BE SILENCED FOR GOOD / THE FLOWERS, UNKINDLED WILL BLAZE ONE LAST TIME AND GO OUT / AND THE BOY NURSING HIS STUNG ANKLE THIS MORNING WILL LOOK BACK AT HIS BRIEF TEARS / WITH SOMETHING LIKE REGRET / REMEMBERING THE AMBER TASTE OF HONEY

ET LATET ET LUCET PHAËTHONTIDE CONDITA GUTTA / UT VIDEATUR APIS NECTARE CLAUSA SUO / DIGNUM TANTORUM PRETIUM TULIT MA LABORUM / CREDIBILE EST IPSAM SIC VOLUISSE MORI

DAYS OF WINE & HONEY

STEAMED ARTICHOKES WITH HONEY DRESSING

HONEY GLAZED CARROTS

GOLD RUSH

HONEY ROAST LEG OF LAMB

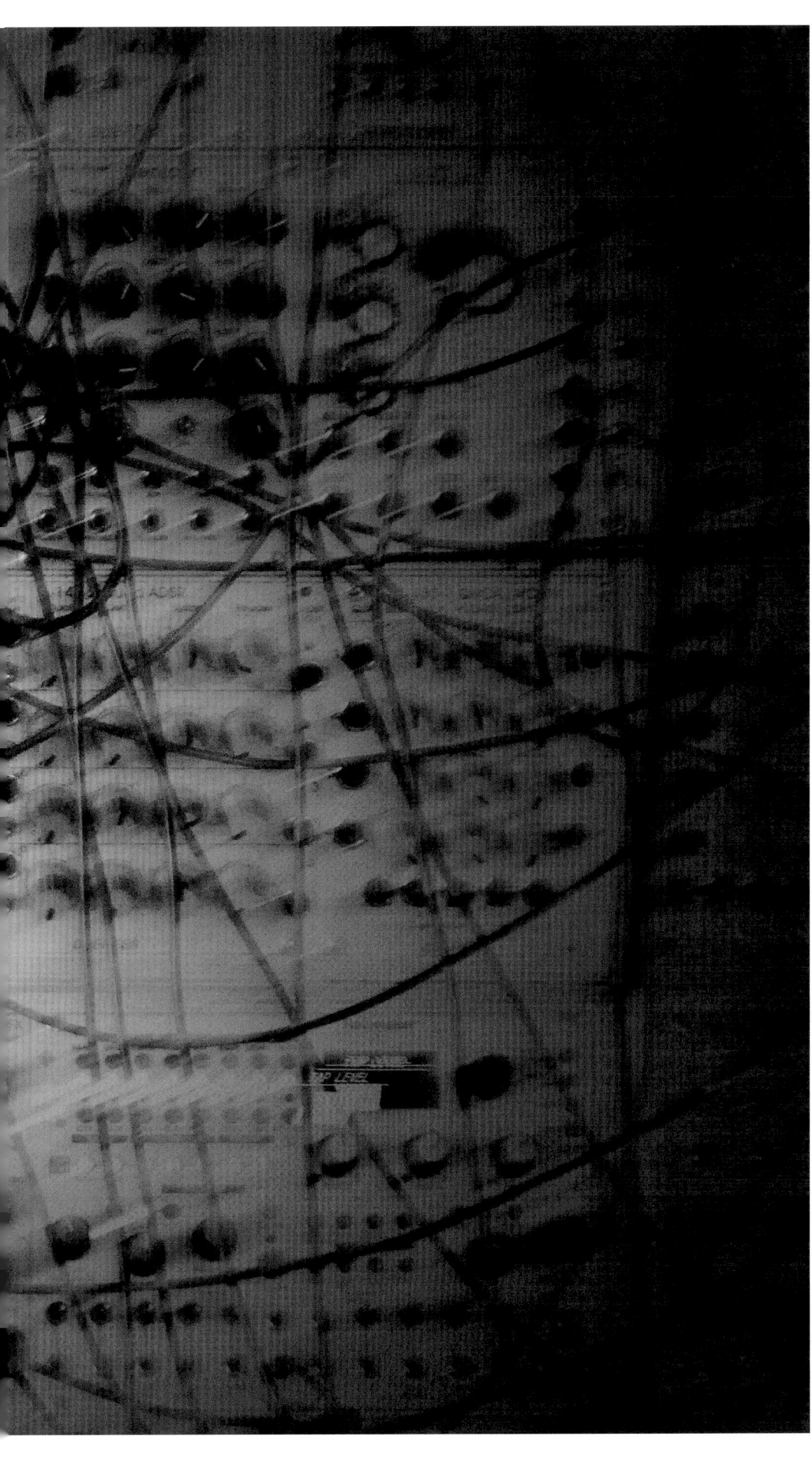

Koen Roggen | Sound Art Work
Download version:
www.imageboulevard.com/
the-poetry-of-the-bee/

WILD REMEDIES BY ANDREW WRIGHT

THE STING IN THE TALE

Many people would go out of their way not to be stung by a bee. Not Andrew Wright. He cherishes the bees he keeps, and more than that: he specialises in apitherapy, using bee venom to cure or treat people. A tale about the sting, told in Malindi, Kenya.

Andrew Wright (68) is a man of few words, and when asked about his love of bees, he simply says: "I was growing anthurium flowers in greenhouses in Malindi in 1996, and one day a swarm of bees flew into my office. I found them so interesting that I bought a beehive."

He explains how he came to apitherapy: "After a few months I had acquired several hives, and I realised that beekeeping could become more than a hobby: I wanted to do it commercially. But I soon found out that selling honey wasn't making me any money, so I started looking into by-products. That's how I came across apitherapy." And although that didn't make him rich either, he did discover a true passion: bee sting therapy.

The use of bee venom to treat ailments is still quite controversial. Every now and then, we hear something about it in the news: Gwyneth Paltrow recommends it as a beauty treatment, Gerard Butler went into anaphylactic shock twice after being treated with bee venom for a muscle injury, and from time to time, some Hollywood celebrity tries to turn it into 'the next big thing'. But many people still consider it as rather off-the-wall, although in fact apitherapy has been around for a very long time. The healing properties of honey and other bee products are mentioned in many religious works, including Vedic texts, the Bible and the Quran. The origins of bee venom therapy can be traced back to Ancient Egypt and Greece and it has been practiced in China for over a thousand years.

Andrew gained a deep insight into the therapy by studying the work of Charles Mraz, pioneer of bee venom therapy in the United States. Mraz himself was inspired by Hungarian bee expert Bodog Beck, who wrote ' Bee Venom Therapy: Bee Venom, Its Nature and Its Effect on Arthritic and Rheumatoid Conditions' (1935), by now a classic in this field. Beck, in turn, learned from the Austrian Dr. Anton Terc who studied the effects of bee venom in the 19th century. In other words: Andrew became part of a long tradition of healers.

According to experts, bee stings put the body's immune system under stress, helping it to grow stronger. Scientific evidence suggests that bee products have healing properties: they improve circulation, calm inflammation and stimulate a healthy immune response. Bee venom contains about forty healing ingredients, including melittin that has anti-inflammatory, anti-arthritic, anti-cancer and anti-microbial properties.

"It was the most interesting adventure I have ever embarked on," says Andrew. "I just wanted to learn everything I could about bee sting therapy." He has now treated more than 3000 people and claims he has helped cancer patients and MS sufferers, alleviated the pain of arthritis and - though it does not cure the disease - even improved the health of Aids patients. Apart from stinging his patients with bee venom, he combines the treatment with the use of other products such as propolis, pollen, beeswax and chitosan.

Andrew Wright now owns some hundred hives, dotted around the Malindi area. He has worked out a 'sharing system' where he places hives on the land of poor farmers. "All they have to do is make sure it isn't stolen, we do the rest and they get their share of the profit."
He is still exploring new ways of working with bees - "I have just tried to get the bees to build their comb in a glass jar and fill it with honey" - and he loves bee paraphernalia: "I have this very old bee book for children, written in the early 1900's. Would you like to see it?" He has a great love for the apis mellifera scutellata, or African bee.

Although he is deeply dedicated to his work, it hasn't always been easy. "Bee therapy is very complex, and in the early days, I often thought of giving up. But when someone comes to you hurting all over, and you sting them with bee venom for a month, gradually increasing the number of stings to 30 a day, and he suddenly tells you that his pain is gone, it's wonderful. Nothing short of miraculous, really." ✖

158 | The natural habitat of the Kenian Killer Bee in Malindi

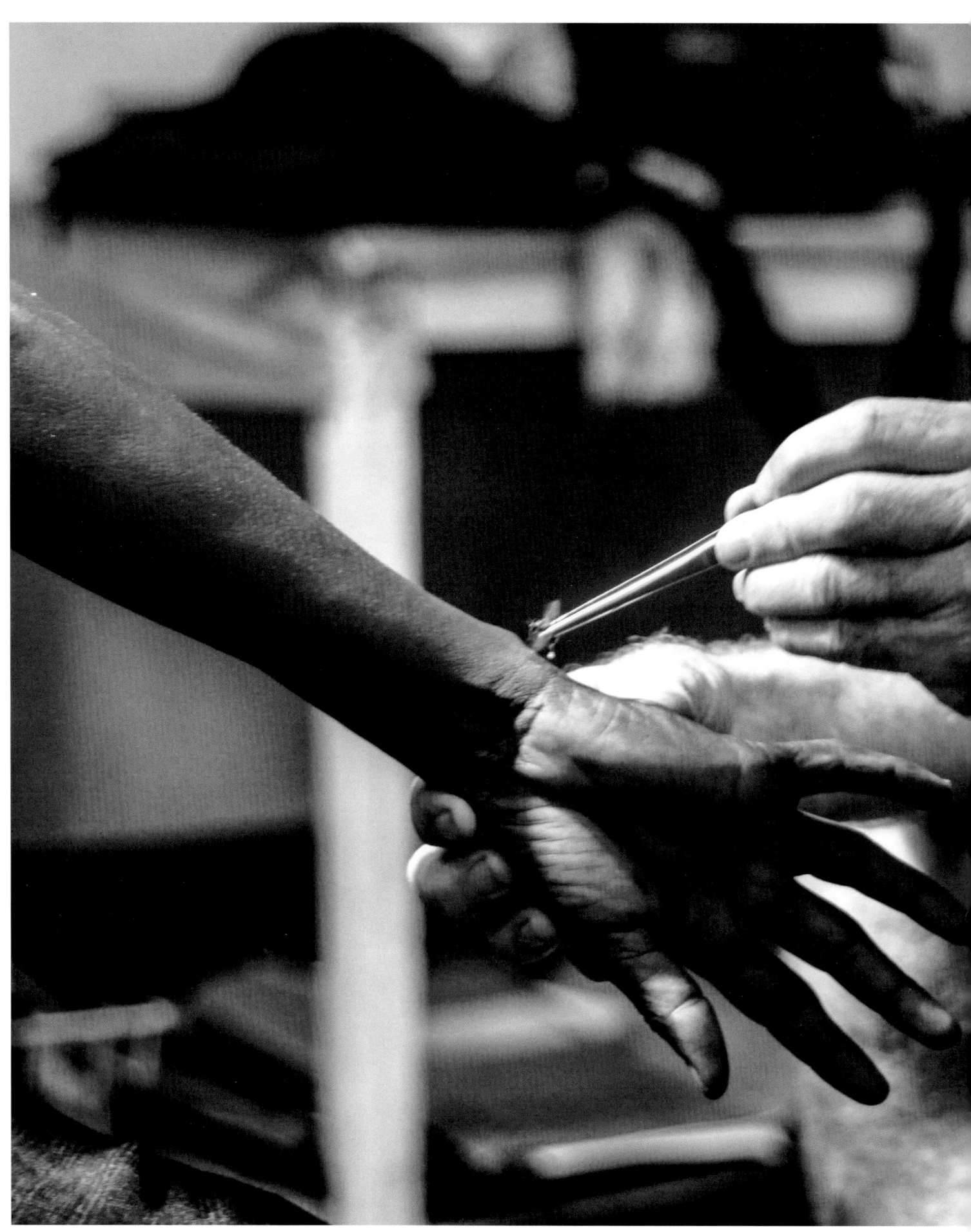

160 | An Aids patient undergoing apitherapy treatment

MASTER BUILDERS

Artist Ren Ri | Yuansu Series II | Acrylic box & natural beeswax

The beehive was the inspiration for this block of flats by Le Corbusier

168 | Frank Lloyd Wright used a full screen of regular hexagons for the ground plan of Hanna House | Stanford California 1936/7

ARE BEES BETTER ARCHITECTS THAN PEOPLE? MAYBEE

Bees are builders. They have to be. As soon as a new fleet is formed, they have to settle. When the bee population (or part of the population) leaves the former hive, scout bees lead the way to a new space for a queen and her brand new colony. Once they arrive at the spot, they immediately start hammering, chiselling, plastering and gluing. It is essential to do this quickly, because the bees squeeze the building material from their own bodies. Those bodies (that were helicopters a few hours earlier) become concrete mixers. As soon as there is enough room for stocking up, forager bees fly out. They, too, have to be quick. It takes 7.5 kilos of honey to make 1.2 kilos of wax (the substance builder bees excrete). In the meantime, eggs need to be laid to provide a new brood. Every bee in the colony becomes a builder in the first or second week of its life: their wax glands are at their most productive between day 12 and day 18. Only after three weeks do they begin to fly out to find nectar. But if needs be, they can still help with the building work.

Wax is very pliable, writes ethologist and bee expert Jürgen Tautz. 'The bees can determine what sort of essential properties they want to produce in their material. It is as if a plasterer sweats out his own bricks and can, at the same time, determine their properties to comply with the specific wishes of the customer.' A honeycomb is a chain of tubes that are originally round. At a certain moment they are warmed up and start to melt, and suddenly they become hexagonal instead of round. This happens because at a certain moment they start to stick together and 'straighten' each other out. This is more or less the final shape. These tubes (cleverly tilted so no honey can be spilled) have all sorts of functions: they store honey, pollen or the brood. The function can also change. Propolis is also used, Tautz tells us: 'The bees scrape it off the plants in the form of resin and use it as a finishing layer on the cell walls, or build it into the walls.'

Bees regulate the temperature in their hive themselves. By moving further apart, but also by bringing in water and letting it evaporate. This is done by special heater bees that are positioned in specific places; they start to tremble and produce warmth. There are also ventilator bees that can flap their wings, in order to regulate the temperature for the brood and the honeycomb, which needs to remain sturdy enough to dance on.

From ancient times, the bees' architecture has been appreciated. According to author Bee Wilson, a text from the 12th century praised it like this: 'What house with four walls shows as much craftsmanship and beauty as the framework of their honeycombs, with small round rooms supporting each other by sticking together? What architect taught the bees to fit hexagonal rooms together when the lengths of the sides are indistinguishable?'

The fact is that bees are masters in certain skills. In building straight without surveyors and spirit levels, for instance. 'The honeycombs hang completely vertically,' author Jurgen Tautz writes. They need to, because that benchmark is needed when bees are dancing on the honeycomb to indicate where food or other targets can be found. 'Without these perpendicular honeycombs this form of communication would never have developed,' says Tautz. 'Insects that form colonies without such perpendicular honeycombs cannot possibly clearly indicate a certain corner by dancing.' It doesn't matter what the branch they hang from looks like, or the rock wall or the hollow tree, or even the prefabricated beekeeper's honeycomb. It is the principle that is also used by Tomáš Libertíny: give bees a shape and they work with it. But they create something that serves them well: something large enough to house the colony and the energy stock.'

In 2000, author Juan Antonio Ramirez described how the great 20th century architects were intrigued by bees in his book 'The Beehive Metaphor, from Gaudí to Le Corbusier' (which can be found online). He gives a detailed description of how architects such as Gaudí, Frank Lloyd Wright, Ludwig Mies van der Rohe and Le Corbusier used the bees' building techniques and/or methods in their own, often radical ideas and theories.

172 | Japanese architect Shigeru Ban used hexagons for the roof of the Centre Pompidou

174 | Ren Ri | Yuansu Series II | Acrylic box & natural beeswax

Up till now, the 21st century has also produced its fair share of bee architecture, but for the time being, it is mainly for aesthetic reasons. The pattern of linked hexagons keeps inspiring architects, not only because of its strength, but also because of its beauty. In 2002, architect Toyo Ito built a pavilion in Bruges, Belgium, consisting of aluminium panels in the shape of honeycombs. The pavilion was removed in 2013, the same year that the Japanese architect received a Pritzker Prize for his work. In the pavilion, the honeycomb was an essential element to allow light in, filter and reflect it. From certain angles, the structure was almost invisible because you could see straight through the honeycombs. The lightness and magic of the place was due to the honeycomb.

The hexagonal cells proved to be not only extremely suitable for storing honey, the honeycomb is also one of the most efficient ways to keep a construction standing. The ancient Greeks knew this already, and even these days, honeycomb constructions are still used in airplanes, buildings, design and packaging. In other words, anywhere where limiting the amount of material used is an asset.

At the moment, BIG, the company of Bjarke Inghels from Denmark, is building a luxury apartment complex in the Bahamas, consisting of elongated hexagons, which they called the Honeycomb. The lowest point of each hexagon contains the deepest point of a... swimming pool. By keeping the lowest point there, precisely above the supporting wall of the underlying hexagon, the weight of the water can be carried by the floating balcony.

And Shigeru Ban, the Japanese architect who likes to keep it light and who also won a Prizker Prize (in 2014), used hexagons for the roof of the Centre Pompidou – Metz in 2010. The inspiration, in this case, didn't come from bees. "With a surface area of 8,000 m2, constructed fully in wood, the roof is made up of hexagonal units resembling the cane-work pattern of a Chinese hat," says the architect, who collaborated with Jean de Gastines. He has another reason to use the hexagon. Two in fact. "To the French, the hexagon is a symbol of their country, as it is similar to the geographical shape of France. By creating a pattern of hexagons and triangles only four wood elements ever intersect. The intersections do not use mechanical metal joints... Instead, each element overlaps, as in bamboo wickerwork. This idea came from a traditional woven Chinese hat I found in an antiques shop in Paris in 1999 while designing the Japanese Pavilion for the Hanover Expo."

The one contemporary architecture project that was inspired by bees more than any other is the British pavilion at the 2015 Milan World Expo, which has received several prizes. The Hive is a feather-light steel building designed by artist Wolfgang Buttress, presented in a sort of garden. Not only is it shaped like an abstract beehive, it is inspired by the vibrations that bees send each other to communicate. Visitors to the pavilion were treated to a visual feast, with LED lights controlled by the vibrations of a real bee colony. After a year, the pavilion moved to Kew Gardens, where it stood until the end of 2017. It gained numerous awards: it was Best Pavilion at the Expo, CNN praised it as one of the most visually inspiring moments of 2016, and in the same year the British Landscape Institute nominated it Best Temporary Landscape Design. Until now, the Hive is the closest thing we have to the experience of a bee hive and the life of the bee. And it was set in the middle of nature, as it should be... ✖

Architect Toyo Ito built a pavilion consisting of aluminium panels in the shape of honeycombs | Bruges, Belgium 2002

The Hive, a feather-light steel building designed by artist Wolfgang Buttress

MICHAEL VAN LENT

CHEERS TO THE NEW GENERATION OF BEEKEEPERS

A new generation of urban beekeepers is appearing all over the world. In Belgium, we have Save the Queen, who keep beehives in various cities and who make their own gin, rum and elderflower liqueur with the honey. Although what it's really all about, is the story of the bee.

Michael Van Lent is the backbone of Save The Queen. He takes us to see the two beehives on the roof of The Post Hotel on the Korenmarkt in the centre of Ghent. As we arrive, the sun shines down on the beehives, and immediately, the buzzing gets louder. "They become more active when they feel the sun," says Michael. He points at the ground: "Look, that wasp is eating a dead bee. Nature is so great." He doesn't see himself as a die-hard environmentalist. "Only when it comes to bees. My family owned a slaughterhouse, which is not very environmentally friendly. And yet I became a beekeeper because of my father. He had always wanted to do something with honey. When I left for Australia six years ago, he said: 'Check out the bee industry there.' Australia doesn't have the varroa mite, so their honey production is big. I got in touch with beekeepers there. The first time you open a hive… you'll never forget it. When I returned to Belgium, I persuaded Olivier - now a partner in Save The Queen - to start keeping bees together. We didn't know what to do with the honey. Gin was just becoming a big hype and we thought: let's make gin from honey, just for fun. We then went on to make rum and elderflower liqueur. It was an instant success. There are more beekeepers who use honey in liquor, but honey sediment in the bottle is something I have never seen before. Some people see it and think there is something wrong with the gin. But in fact it's not normal if there is no sediment. Once you remove it, you have to start using flavour enhancers.
Our dream is to start a worldwide community of beekeepers with Save the Queen products, always locally produced. We already work together with two beekeepers in Amsterdam. They have honey, but not the other products. We can provide those.
Many people ask why we don't sell honey. Raw, cold-pressed honey is very popular at the moment. But we need all our honey for our liquor. We have fourteen hives in Ghent and Aalst. In Liege, Antwerp, Brussels, Bruges and Knokke we work with local beekeepers. It wouldn't be a good idea to put more hives in each city: when there are too many hives, the bees don't produce enough honey."

Yes, bees thrive in cities. "That's right," says Michael, "bees suffer in the countryside, because of monoculture, pesticides and artificial fertilisers. In the countryside, the beekeepers lose half their colonies to Colony Collapse Disorder, while in the city only 5% are affected. Pesticides disturb the bees' orientation system, they can't find their way back, so that less food is brought back to the hive, and the colony becomes weaker."

There is another reason why bees become town-dwellers. "In the city there is also a greater diversity of flowering trees and plants. In the countryside you have to find out: when do the surrounding fields flower? In the city something is always flowering. It is warmer as well there, so everything flowers for longer. In the countryside, it is over for the bees by July. Here, I can harvest in August. And because of diversity, the honey has a more complex flavour. Our honey comes from many different flowers. And every kind of honey tastes different, because one hive is nearer to more lime trees, another one closer to more flowers. Our gin tastes different in every town. Also, cities are constantly becoming greener: low emission zones, the council using fewer pesticides… Did you know that Brussels is one of the greenest cities in Europe? There is the Sonian Forest, and the suburbs are full of green. You'd think that the honey would contain a lot of fine dust, lead or diesel, but they are filtered out by the plants and the bees. Urban honey is four times purer than honey from the countryside. Last year at the Apimondia Cultural Congress, Martin Liot from Paris won the prize for the world's second best honey, that says it all."

Urban beekeepers are the future, according to him. "Although older beekeepers won't like to hear it. Traditional beekeepers can be rather pig-headed. They each have their own theory. When I just started, I hardly got any tips from them. I get far more now. I look up a lot, I have a mentor from De Vlijtige Bie in Aalst, I go to meetings. They want to know that you're serious about it. But urban beekeepers are everywhere: Paris, Geneva, London, Berlin, Seoul… And there are so many possibilities: collaborations with museums, historic buildings, rooftop farms, bars… David Lebeer, barman at The Cobbler, won the Belgian finals of the Diageo World Class competition with a cocktail that contained our honey, harvested on the roof. How great is that? David adores bees, he even has a bee tattoo. We couldn't wish for a better ambassador. It is hard work, but urban honey is up-and-coming. We are happy about this. And so are the bees. Bees like city life. Win-win."

Yet to him, the bees are the most important. "We want to use our products to make people think of what they can do for the bees. Twice a week, we hold an introductory day for our partners, all about bees. I've had a talk at Guerlain. They have products with royal jelly and they protect the black bees that only live on an island in the north of France: because of their isolated circumstances, they are still

186 | The rooftop of The Post Hotel in Ghent, one of the locations of the beehives

healthy, without any illnesses. It is our dream to be hired for durable initiatives all over the world. Never mind the gin, as long as Save The Queen can continue."

It's a great hype nowadays: everybody wants to save the bees. "We get a lot of requests for beehives, but you have to know what you're doing. People impulsively install a hive and they want bees straight away, while it is better to be patient for a year. If you don't look after your hive, the bees go wild and start swarming. The swarm then often dies. Bees are no longer used to making a nest in the wild. It is sad, but a bee colony can no longer live on its own. You have to look after your bees." Beekeeping has become quite fashionable, but Michael assures us that it does take a lot more work than just installing a hive, letting the bees do their work and collecting the honey. "You have to check if there is illness in the hive, if the eggs have hatched well, or if there is enough honey for the winter. It is hard, specialised work to look after bees. It would be sad for the bees if everybody installed a beehive and then found it too hard, and if people only used beekeeping as something to boast about at the next garden party."

He concludes: "Anyway, we won't save the world by installing more beehives alone. There would then be too many honeybees compared to wild bees - who have a hard time already. No, we have to bring back nature, and return to healthier agriculture with fewer pesticides. Fewer woods should be chopped down, because trees are important to the bees, to make propolis from the resin, they need nectar from the fruit trees, lime trees, acacias, willows... People can help by sowing flowers in the meadows, or by letting their garden go wild. If we leave nature alone, it can still turn out alright. How many bees does it take to save the world? I ask: how many flowers and trees does it take?" ✖

ECHINOPS SPHAEROCEPHALUS

BLOSSOM TIME

ANEMONE CORONARIA

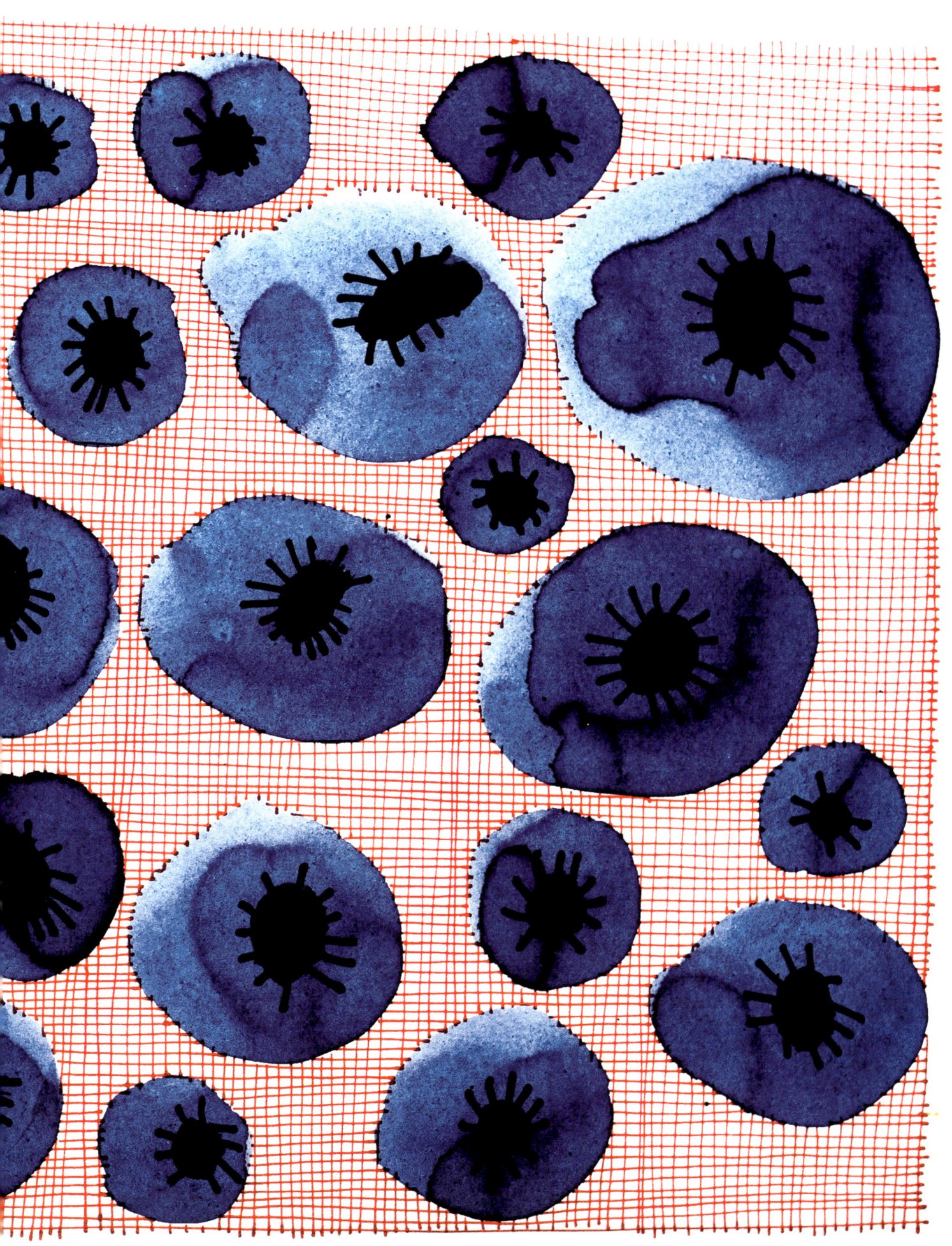

ASTER LAEVIS

RECIPES DAYS OF WINE & HONEY

Gold rush

For 1 cocktail: 30 ml whiskey • 1 tbsp freshly squeezed lemon juice • 120 ml apple cider • 2 tbsp honey syrup • ice cubes

Method: pour all the ingredients into a shaker and fill with ice cubes. Shake for 10 seconds and pour the cocktail into a whiskey glass filled with ice.

For the honey syrup, heat 185 ml honey with 185 ml water and 3 cinnamon sticks, stirring continuously. The mixture should not come to the boil. Leave to cool. Store in a tightly sealed bottle in a dark, cool place.

Steamed artichokes with honey dressing

Serves 4: 4 artichokes • 3 tbsp acacia honey • 4 tbsp apple vinegar • 3 tsp mustard • 50 ml olive oil • pepper and salt

Method: rinse the artichokes under a cold tap. Cut off the stem as closely as possible to the base and remove any furry bits. Steam the vegetables about 45 minutes until tender. Whisk the ingredients for the dressing until smooth. Add pepper and salt to taste. Sprinkle the artichokes with some olive oil and serve with the honey dressing.

Honey roast leg of lamb

Serves 4: 1 leg of lamb (1.2 kg) • 8 garlic cloves, crushed • the grated rind of 1 orange • 4 sprigs rosemary • 180 ml orange blossom honey • 90 g brown sugar • 125 ml white wine • 125 ml orange juice • 1 l lamb stock • olive oil • pepper and salt

Method: preheat the oven to 180°C. Season the lamb with pepper and salt and sear in hot olive oil until golden brown on all sides. Put the meat in a roasting tin. Bring the stock to the boil with the garlic, orange juice and zest, rosemary, brown sugar, honey and white wine. Pour over the leg of lamb, cover in foil and roast in the oven for 2.5 hours. Increase the oven temperature to 220°C, remove the foil and roast the meat for another 45 minutes. Turn the roast regularly during cooking.

Honey glazed carrots

Serves 4: 2 bunches rainbow carrots • 6 tbsp thyme honey • 1 tsp garlic powder • 1 tbsp fresh thyme leaves • 9 tbsp butter • pepper and salt

Method: preheat the oven to 180°C. Scrape the carrots. Grease an oven proof dish and place the carrots in it. Melt the butter, stirring continuously. Take the pan off the heat and stir the honey, garlic powder and thyme into the butter. Season with pepper and salt. Pour over the carrots and cook them in the oven for 30 minutes, turning them regularly.

AND FINALLY...

Two years ago, when I thought of making a book about the honeybee, I had no idea of the beauty I would come across. After 20 years of working as an art director for various publishers, I felt a strong need to work out a beautiful concept that would not only look stylish, but that would also have a social value. It was to be on a subject or theme that did not stay on the surface, but would go deeper. The more I began to read about Apis Mellifera, the more fascinated I became. The honeybee would become the theme of a large format book with strong photography, art and lifestyle. I wanted to create a hive of writers, journalists, artists and all sorts of creative people fascinated by the bee.

There are plenty of scientific books about bees, but I intended to introduce the reader to the world of the bee in a surprising way. With beautiful, well thought-out images and writing, I wanted to celebrate an insect that is not just any insect, but that has an enormous influence on our eating patterns. Without pollination we wouldn't have vegetables to eat, as Nobel Prize winner Maurice Maeterlinck concluded 100 years ago.

I intended to highlight the new evolution in beekeeping, and that is why we looked at five very different beekeepers and their individual approach to their bee colonies. I also wanted to look into apitherapy, remedies with bee venom, because it has become clear that the bee as a pharmacist still has a large role to play. I visited beekeeper Andrew Wright in Malindi, Kenya and was deeply impressed with the wonderful results he obtains using bee stings as medicine.

Just as impressive were the artists I came to know during my research. Many of them have been inspired by the honeybee, and they all have their own approach. From the beeswax sculptures of Tomáš Libertiny and the Honey Pump of Joseph Beuys to Jan Fabre's Bic Art and the wonderful installations of duo Carla Arocha and Stéphane Schraenen, they all seemed to be just right for this book. I was also fascinated by the work of Le Corbusier and found other architects who were inspired by the clever bee population.

Nowadays, bees like to live in the city. We heard it from beekeepers and we saw it in Manchester, the city that chose the bee for its symbol. We went on a walking tour there and discovered how the bee is part of the ancient and recent history of this British city.

I could have made a book as fat as a bible, but I had to limit myself to 208 pages. 'Kill your darlings', it is called: make difficult choices, and leave things out with pain in your heart. But I hope that my selection gives a different view of the wonderful world of the honeybee. I thank everybody who has taken part in the production of this book. Luckily, nobody got stung during the photo shoots and interviews. And please; don't forget to plant flowers for the bees.

Iris Rombouts has 20 years of experience in the world of publishing. During those years, she has put her stamp on the Flemish media landscape as a designer of upscale magazines. Over the past nine years, she has created concepts and projects for her company Image Boulevard, specialising in art direction, photography and graphic design for art, fashion and lifestyle publications. In 2014 she created digital platform 'Les Petits Belges', an influencer magazine about art, fashion, architecture and lifestyle.

CREDITS

Cover
Lasercut | Carla Arocha
and Stéphane Schraenen
Photography | Stephen Mattues -
Filip Van Roe - Joost Joossen

Maurice Maeterlinck
Words | Tony Rombouts
Photography | Stephen Mattues

Landscapes
Multi Media | Stephen Mattues
& Iris Rombouts

Louis De Cordier
Interview | Lene Kemps
Photography | Stephen Mattues
www.cosco.one

The Beauty of Pollination
Words | Felix Wäckers
Photography | Felix Wäckers

Jan Fabre
Words | Anne Davis
Photography | Stefan Vanfleteren

Hoeder zelfportret I: 1991
Material: ballpoint on colour photo
Format: 146 x 207 cm
Photography: Studio Ghezzi
Collection: private collection, Marc
Vandijck & Laure Vandersmissen
© Angelos bvba
Hoeder zelfportret II: 1991
Material: ballpoint on colour photo
Format: 146 x 207 cm
Installation: Jan Fabre. Messengers
of the Death.
2004 Salzburg, Galerie Mario
Mauroner Contemporary Art
(Jul 2004 - Sep 2004)
Photography: Studio Ghezzi
Collection: Privécollectie, Marc
Vandijck & Laure Vandersmissen
© Angelos bvba
Imker (II) (van 1998 tot 1999)
Material: jewel beetle on wire
Format: 80 x 75 x 160 cm
Installation: Jan Fabre. Homo Faber.
(12 Mei 2006 - 3 Sep 2006)

Photography: Attilio Maranzano
Collection: private collection
© Angelos bvba

Carla Arocha & Stéphane Schraenen
Words | Iris Rombouts
Photography | Jean Pierre Stoop

Beekeeper Fashion
Interview | Gerdi Esch
Photography | Stephen Mattues
Photo assistant | Simon Waterkeyn
Styling & coördination |
Ellen Monstrey
Hair & make-up | Rudi Cremers
Models | Femke - Paparazzi Models
& Stephan Van De Velde - Ys Models
Van zwolharm@hotmail.com
Special thanks to Academie
Sint-Niklaas Belgium
Sint-Ambrosius Imkersgilde Edegem

Tomáš Libertiny
Words | Leen Creve

Seed of Narcissus
Material: beeswax, glass, 2011
Dimensions: 950 x 400 x 400 mm
Edition: unique
Collection: MUDAC Museum
of Lausanne ade in collaboration with
Berengo Studios for Glasstress 2011

Landscape
Material: beeswax, wood, 2013
Dimensions: 590 x 590 x 500 mm

Joseph Beuys
Words | Jody Stokes Casey
Images | Licensing Visual arts
© SABAM Belgium 2018

Manchester city of bees
Interview | Anne Davis
Photography | Filip Van Roe

Beautiful honey body
Words | Sofie Albrecht
Photography | Stephen Mattues
Sources:
• BBC programme Hive Alive,
broadcast on 22-7-2014

• The Honey Prescription,
by Nathaniel Altman,
published by Healing Arts Press.
• Why super honey is the bees'
knees for wounds and infections,
published by 01-01-2014 n
www.theguardian.com
• Le miel: un cicatrisant naturel à
l'hopital, published on 25-07-2017
on www.nationalgeographic.fr
• Mon médecin est une abeille,
published on 10-4-2009 in Le Point

Days of Wine and Honey
Words | Sonja Peeters
Photography | Wout Hendrickx

**Untitledvariousunknown
beesconcertonr.01**
Sound Art Work | Koen Roggen
Photography | Koen Roggen

Andrew Wright
Words| Lene Kemps
Photography | Iris Rombouts

Master Builders
Words | Leen Creve
Licensing | Centre Pompidou-Metz,
2010 | © Shigeru Ban Architects
Europe and Jean de Gastines
Architects, with Philip Gumuchdjian
for the design of the winning project /
Metz Métropole / Center Pompidou-
Metz / Photo Philippe Gisselbrecht
Photography | studioEAST - Ren Ri -
Getty Images - Deposit Photo - Mark
Haddon

Michael Van Lent
Words | Ineke van Nieuwenhove
Photography | Diego Franssens
www.savethequeen.eu

Blossom Time
Illustration | Ann Rikkers

Contact information:
Iris Rombouts
creative@imageboulevard.com

Special thanks to: Casstl, Tom Van Noten, Peter Ruyffelaere, Johan Van Roy, Stephen Mattues, Anne Davis, Lene Kemps and all the contributors who helped this project come together.

COLOFON

CONTENT - ART DIRECTION - GRAPHIC DESIGN
IRIS ROMBOUTS

COPYWRITER | ANNE DAVIS

PUBLISHER | ANDER-ZIJDS | 2018

[ander]-zijds

ISBN 9789082808025
NUR 640
D/2018/14.359/03

DUTCH DOWNLOAD VERSION: WWW.IMAGEBOULEVARD.COM/THE-POETRY-OF-THE-BEE/